The Big Picture

The Universe in Five S.T.E.P.S.

Synthesis Lectures on Engineering, Science, and Technology

The Big Picture: The Universe in Five S.T.E.P.S.
John Beaver
2020

Relativistic Classical Mechanics and Electrodynamics
Martin Land and Lawrence P. Horwitz
2019

The Big Picture: The Universe in Five S.T.E.P.S.

John Beaver

ISBN: 978-3-031-00952-4   paperback
ISBN: 978-3-031-02080-3   ebook
ISBN: 978-3-031-03208-0   hardcover

DOI 10.1007/978-3-031-02080-3

A Publication in the Springer series
SYNTHESIS LECTURES ON ENGINEERING, SCIENCE, AND TECHNOLOGY

Lecture #2
Series ISSN
ISSN pending.

The Big Picture

The Universe in Five S.T.E.P.S.

John Beaver

University of Wisconsin Oshkosh

SYNTHESIS LECTURES ON ENGINEERING, SCIENCE, AND TECHNOLOGY #2

ABSTRACT

A brief overview of astronomy and cosmology is presented in five different ways, through the lenses of space, time, evolution, process, and structure. Specific topics are chosen for their contribution to a "big picture" understanding of the interconnectedness of knowledge in astronomy and cosmology. Thus, many topics (stellar astronomy for example) are treated in multiple sections, but from different viewpoints—for example, sizes and distances of stars (space); when stars appeared in the history of the universe (time); stellar evolution (evolution); hydrostatic equilibrium and stellar spectra (process); and stellar structure (structure). Some topics traditional to the introductory astronomy curriculum—eclipses and lunar phases, for example—are omitted altogether as they are inessential for the big-picture goals of the book, and excellent summaries are easily available elsewhere. On the other hand, the book treats some topics not usually covered in an introductory astronomy course, for example the roles played by equilibrium processes and symmetry in our understanding of the universe. The level is for the beginning undergraduate, with only basic skills in rudimentary algebra assumed. But more advanced students and teachers will also find the book useful as both a set of practical tools and a point of departure for taking stock (in five different ways) of the current state of knowledge in astronomy and cosmology.

KEYWORDS

astronomy, cosmology, evolution, stellar spectra, hydrostatic equilibrium

For Doug

Contents

Preface

This is a little book about a big subject, and it began as an art exhibit. *Astronomy: Images, Ideas and Perspectives*, a collaborative effort between Judith Baker Waller (Professor of Art at the Fox Cities campus of University of Wisconsin Oshkosh) and myself, premiered in Waupaca, WI in 2000. As part of the preparation for the exhibition, I sat in on Judith's art history class. But I also began writing—for her—a description of key concepts in astronomy and cosmology. My goal was not so much to write a text that would be understandable to an artist. Rather, I wanted my text to be *useful* for Judith's making of art. And so I hit upon the idea of describing astronomy in terms of grand themes—space, time, evolution, etc. As I began to synthesize basic astronomy in these terms, I realized that *I* had much to learn, and that this might be a useful perspective for anyone.

I have divided the material in this book into five complementary and parallel tracks. Each can be seen as a brief description of astronomy and cosmology as a whole, but from only one limited perspective. To fully understand one track, one must understand the others; but each track is written to be as stand-alone as possible. These five *STEPS* are as follows.

1. **Space:** What are the sizes of, and the distances between, the various things that make up our universe? As an example, how does the size of the Earth compare to its distance from the sun? And how does this compare to the size of the sun, a fairly typical star?

2. **Time:** How has the universe as a whole changed, and what is its future? As an example of just one part of this history, when—in the overall scheme of things—did stars like the sun first begin to form?

3. **Evolution:** How do different parts of the universe change? For example, how does an individual star like the sun change over time?

4. **Process:** What are the basic building blocks and rules of interaction that govern what goes on in the universe? For example, what are the forces that keep a star like the sun round?

5. **Structure:** What makes up what? For example, what larger structures can a star be part of?

Many topics are covered in *each* of these tracks; you will read about stars, for example, in each of the five sections. But in each track you will read about stars from a different perspective. The same holds for galaxies, planets, supernovae, and clouds of interstellar gas. The goal is that the reader will gradually synthesize the material into a "big picture," building an image of not only the different parts that make up our universe, but also how they fit together and interact with each other.

The Big Picture can be used as the foundation for an introductory undergraduate survey of astronomy course. But as the *sole* basis for such a course this book is incomplete, and intentionally so. The reader will find here, for example, no descriptions of eclipses or phases of the moon. And so *The Big Picture* is not intended to be comprehensive; rather, I have included topics only insofar as they directly contribute to an understanding of the five *STEPS*. It is my contention, however, that most of the obviously left-out topics are perfect for student assignments or projects; excellent surveys of eclipses, for example, are widely available online.

John Beaver
December 2019

Acknowledgments

I thank Valeria Sapiain for all-around support and patience over the year of intense work on this book. Many conversations and projects with Doug Fowler, Judith Baker Waller, and Teresa Patrick, over many years, have had a profound influence on me in ways I only partially understand. The concept for and structure of *The Big Picture* were surely influenced by both *Cosmic View: The Universe in 40 Jumps* by Kees Boeke [Boeke, 1957], which I read as a child, and *Coming of Age in the Milky Way* by Timothy Ferris [Ferris, 1988], which I read in graduate school while avoiding dissertation work. The following software packages were used to make many of the illustrations in this book:

GIMP (www.gimp.org) Gnuplot (www.gnuplot.info)
Inkscape (www.inkscape.org) SciDAVis (scidavis.sourceforge.net)
Aladin Sky Atlas (aladin.u-strasbg.fr/) IRAF (iraf.noao.edu)
Stellarium (stellarium.org/) Cybersky (www.cybersky.com)
Google Earth (www.google.com/earth/)

Cover photograph: *Dime Eclipse*, by John Beaver (2017). All photographs and illustrations are by the author, except as noted below and in the individual figure captions:

Figure 1.2: Beaver image by Steve from Washington, DC. USA, commons.wikimedia.org/w/index.php?curid=3963858, CC BY-SA 2.0.

Figure 1.6: Illustration by Duckysmokton, Ilia, Vermeer, commons.wikimedia.org/w/index.php?curid=2888434, CC BY-SA 3.0.

Figure 1.8: Graphic commons.wikimedia.org/w/index.php?curid=1903952, Public Domain.

Figure 2.1: Left-hand image made with Stellarium Stellarium, available from stellarium.org.

Figure 2.2: Image made with Google Earth, copyright 2018 by Google, U.S. Department of State Geographer, image Landsat/Copernicus, Data SIO, NOAA, US Navy, NGA, GEBCO.

Figure 2.3: Images made with Google Earth, copyright 2018 by Google, U.S. Department of State Geographer, image Landsat/Copernicus, Data SIO, NOAA, US Navy, NGA, GEBCO.

Figure 2.6: Image made with Google Earth, copyright 2018 by Google, U.S. Department of State Geographer, image Landsat/Copernicus, Data SIO, NOAA, US Navy, NGA, GEBCO.

Figure 2.7: Illustration by cmglee, David Monniaux, commons.wikimedia.org/w/index.php?curid=52681377, CC BY-SA 4.0.

Figure 2.8: Sky map and planet tracks created with Cybersky, available at www.cybersky.com; used with permission.

Figure 2.11: Image by Rawastrodata, rawastrodata.com/dso.php?type=globularclusters&id=m13, CC BY-SA 3.0.

Figure 2.12: Right-hand image made with the Aladin Sky Atlas. DSS2, alasky.u-strasbg.fr/DSS/DSSColor [Bonnarel et al., 2000, Lasker et al., 1996].

Figure 2.14: Image by ESO/S.Brunier www.eso.org/public/images/eso0932a/, CC BY 4.0.

Figure 2.15: Image by Hewholooks - Own work, commons.wikimedia.org/w/index.php?curid=4290900, CC BY-SA 3.0.

Figure 2.16: Adam Evans, commons.wikimedia.org/w/index.php?curid=12654493, CC BY 2.0.

Figure 2.18: Image credit: NASA/STScI/WikiSky, commons.wikimedia.org/w/index.php?curid=7598267, Public Domain.

Figure 2.20: Special thanks to Olivia, Phoebe, Sophie and Emily.

Figure 2.21: www.spacetelescope.org/images/heic0611b/ NASA, ESA, and S. Beckwith (STScI) and the HUDF Team, Public Domain.

Figure 2.22: NASA/WMAP Science Team. commons.wikimedia.org/w/index.php?curid=3753192, Public Domain.

Figure 5.1: Graphic by Brews ohare, commons.wikimedia.org/w/index.php?curid=6042242, CC BY-SA 3.0.

Figure 5.2: Graphic by Kintpuash, commons.wikimedia.org/w/index.php?curid=68991059, CC0.

Figure 5.3: Graphic by Quantum Doughnut - Own work, commons.wikimedia.org/w/index.php?curid=12958270, Public Domain.

Figure 5.4: Graphic by Swift - Own work, commons.wikimedia.org/w/index.php?curid=48991521, CC0.

Figure 6.2: Graphic, adaptation of original NASA WMAP Science Team image lambda.gsfc.nasa.gov/education/graphic_history/univ_evol.cfm.

Figure 7.1: Graphic by BenRG - Own work, commons.wikimedia.org/w/index.php?curid=7881135, Public Domain.

Figure 8.1: Images by NASA, Mercury image: JHUAPL Venus image: JPL Mars image: HST; Mercury Globe-MESSENGER, commons.wikimedia.org/w/index.php?curid=39782734, Public Domain.

Figure 8.2: Graphic by Lunar and Planetary Institute solarsystem.nasa.gov/galleries/gas-giant-interiors, Public Domain.

Figure 8.3: NASA/JPL-Caltech, www.jpl.nasa.gov/news/news.php?feature=7194.

Figure 8.4: NASA, nssdc.gsfc.nasa.gov/imgcat/html/object_page/nea_19970627_mos.html, Public Domain.

Figure 8.6: NASA/JPL-Caltech/UMD photojournal.jpl.nasa.gov/catalog/PIA02127, Public Domain.

Figure 8.8: ALMA (ESO/NAOJ/NRAO), commons.wikimedia.org/w/index.php?curid=36643860, CC BY 4.0.

Figure 9.1: ESO, www.eso.org/public/images/eso0728c/, CC BY 4.0.

Figure 9.2 – Figure 9.4: NASA, ESA, M. Robberto (Space Telescope Science Institute/ESA) and the Hubble Space Telescope Orion Treasury Project Team, commons.wikimedia.org/w/index.php?curid=1164360, Public Domain.

Figure 9.5: Graphic by Lithopsian - Own work, commons.wikimedia.org/w/index.php?curid=48486177, CC BY-SA 4.0.

Figure 9.6: Graphic by Szczureq - Own work, commons.wikimedia.org/w/index.php?curid=34794215, CC BY-SA 4.0.

Figure 9.7: Hubble Heritage Team (AURA/STScI/NASA), hubblesite.org/newscenter/archive/releases/1999/01/image/a/, Public Domain.

Figure 9.8: Graphic by Lithopsian, https://creativecommons.org/licenses/by-sa/3.0, CC BY-SA 3.0.

Figure 9.9: NASA, ESA, J. Hester and A. Loll (Arizona State University), commons.wikimedia.org/w/index.php?curid=516106, Public Domain.

Figure 9.10: Graphic by User:Worldtraveller - Own work, commons.wikimedia.org/w/index.php?curid=1336088, CC BY-SA 3.0.

Figure 11.1: Henry Cavendish, "Experiments to determine the Density of the Earth" in McKenzie, A.S. ed. Scientific Memoirs Vol.9: The Laws of Gravitation, American Book Co. 1900. commons.wikimedia.org/w/index.php?curid=2621520, Public Domain.

Figure 11.2: Earth image: NASA/Apollo 17 crew; taken by either Harrison Schmitt or Ron Evans, commons.wikimedia.org/w/index.php?curid=43894484, Public Domain. Beaver image by Steve from Washington, DC, USA - American Beaver, commons.wikimedia.org/w/index.php?curid=3963858, CC BY-SA 2.0.

Figure 11.5: ESA/Hubble & NASA, www.spacetelescope.org/images/potw1151a/, CC BY-SA 4.0.

Figure 11.6: EHT Collaboration, www.eso.org/public/images/eso1907a/ (image link). The highest-quality image (7416x4320 pixels, TIF, 16-bit, 180 Mb), CC BY 4.0.

Figure 12.4: By 4C - Own work, commons.wikimedia.org/w/index.php?curid=1017820, CC BY-SA 3.0.

Figure 12.5: Image generated by the star map software Cybersky (available at www.cybersky.com); used with permission.

Figure 12.6: Data by BAA (J. Montier), britastro.org/specdb/index.php.

Figure 12.7: Data by BAA (T. Rodda), britastro.org/specdb/index.php.

Figure 12.9: Image made with the Aladin Sky Atlas. DSS2, alasky.u-strasbg.fr/DSS/DSSColor [Bonnarel et al., 2000, Lasker et al., 1996].

Figure 12.10: Images made with the Aladin Sky Atlas. DSS2: alasky.u-strasbg.fr/DSS/DSSColor [Bonnarel et al., 2000, Lasker et al., 1996]. KMASS: University of Massachusetts & IPAC/Caltech [Skrutskie et al., 2006]; AKARI: University of Tokyo, ISAS/JAXA, Tohoku University, University of Tsukuba, RAL and Open University [Doi et al., 2015].

Figure 12.11: Image made with the Aladin Sky Atlas. DSS2, alasky.u-strasbg.fr/DSS/DSSColor [Bonnarel et al., 2000, Lasker et al., 1996].

Figure 12.12: By 4C - Own work, CC BY-SA 3.0, commons.wikimedia.org/w/index.php?curid=1017820.

Figure 13.1: By Wersje rastrowa wykonal uzytkownik polskiego projektu wikipedii: Artura Jana Fijalkowskiego (WarX), Zwektoryzowal: Krzysztof Zajaczkowski, commons.wikimedia.org/w/index.php?curid=11215685, GFDL.

Figure 13.2: NOAO/AURA/NSF, www.noao.edu/image_gallery/html/im0649.html, Public Domain.

Figure 14.1: Graphic by Tomruen, commons.wikimedia.org/w/index.php?curid=56702258, CC BY-SA 4.0.

Figure 14.2: Oort Cloud: ESO/L. Calçada www.eso.org/public/images/eso1614b/, CC BY-SA 4.0. M13: Rawastrodata, commons.wikimedia.org/w/index.php?curid=24968545. M87: MASA. STSCi, Wikisky, commons.wikimedia.org/w/index.php?curid=7598267, Public Domain.

Figure 14.3: Jupiter: NASA, ESA, Michael Wong (Space Telescope Science Institute, Baltimore, MD), H. B. Hammel (Space Science Institute, Boulder, CO), and the Jupiter Impact Team, www.spacetelescope.org/images/heic0910q/; Proto-planetary Disk: NASA, origins.jpl.nasa.gov/stars-planets/ra4.html, Public Domain; M 31: Adam Evan, commons.wikimedia.org/w/index.php?curid=12654493, CC BY 2.0.

Figure 14.4: Graphic by MissMJ - Own work by uploader, PBS NOVA [1], Fermilab, Office of Science, United States Department of Energy, Particle Data Group, commons.wikimedia.org/w/index.php?curid=4286964, Public Domain.

Figure 15.1: Graphic by www.sun.org, www.sun.org/encyclopedia/stars#Structure, CC BY-SA 3.0.

Figure 15.2: Graphics by Marshall Space Flight Center, NASA/MSFC Hathaway, solar-science.msfc.nasa.gov/interior.shtml, Public Domain.

Figure 15.3: R.J. Hall, commons.wikimedia.org/w/index.php?curid=2565862, CC BY 2.5.

Figure 16.1: Images made with the Aladin Sky Atlas. DSS2, alasky.u-strasbg.fr/DSS/DSSColor [Bonnarel et al., 2000, Lasker et al., 1996].

Figure 16.2: Images made with the Aladin Sky Atlas. DSS2, alasky.u-strasbg.fr/DSS/DSSColor [Bonnarel et al., 2000, Lasker et al., 1996].

Figure 16.3: Graphic credit: NASA and ESA. www.spacetelescope.org/images/heic9902o/, CC BY-SA 4.0.

Figure 16.4: Images made with the Aladin Sky Atlas. DSS2, alasky.u-strasbg.fr/DSS/DSSColor [Bonnarel et al., 2000, Lasker et al., 1996].

Figure 16.5: Images made with the Aladin Sky Atlas. DSS2, alasky.u-strasbg.fr/DSS/DSSColor [Bonnarel et al., 2000, Lasker et al., 1996].

Figure 17.1: Willem Schaap, commons.wikimedia.org/w/index.php?curid=7025644, CC BY-SA 3.0.

Figure 17.2: Simulations performed at the National Center for Supercomputer Applications by Andrey Kravtsov (The University of Chicago) and Anatoly Klypin (New Mexico State University). Visualizations by Andrey Kravtsov.cosmicweb.uchicago.edu/filaments.html.

Figure 17.3: Beaver graphic: Public Domain.

REFERENCES

Kees Boeke. *Cosmic View: The Universe in 40 Jumps*. John Day Company, New York, 1957. xix

F. Bonnarel, P. Fernique, O. Bienaymé, D. Egret, F. Genova, M. Louys, F. Ochsenbein, M. Wenger, and J. G. Bartlett. The ALADIN interactive sky atlas. A reference tool for identification of astronomical sources. *Astronomy and Astrophysics Supplement*, 143:33–40, April 2000. DOI: 10.1051/aas:2000331 xx

Yasuo Doi, Satoshi Takita, Takafumi Ootsubo, Ko Arimatsu, Masahiro Tanaka, Yoshimi Kitamura, Mitsunobu Kawada, Shuji Matsuura, Takao Nakagawa, Takahiro Morishima, Makoto Hattori, Shinya Komugi, Glenn J. White, Norio Ikeda, Daisuke Kato, Yuji Chinone, Mireya Etxaluze, and Elysandra F. Cypriano. The AKARI far-infrared all-sky survey maps. *Publications of the Astronomical Society of Japan*, 67(3):50, June 2015. DOI: 10.1093/pasj/psv022 xxii

T. Ferris. *Coming of Age in the Milky Way*. William Morrow and Company, New York, 1988. xix

B. M. Lasker, J. Doggett, B. McLean, C. Sturch, S. Djorgovski, R. R. de Carvalho, and I. N. Reid. The Palomar—ST ScI Digitized Sky Survey (POSS–II): Preliminary Data Availability. In G. H. Jacoby and J. Barnes, Eds., *Astronomical Data Analysis Software and Systems V*, volume 101 of *Astronomical Society of the Pacific Conference Series*, p. 88, 1996. xx

M. F. Skrutskie, R. M. Cutri, R. Stiening, M. D. Weinberg, S. Schneider, J. M. Carpenter, C. Beichman, R. Capps, T. Chester, J. Elias, J. Huchra, J. Liebert, C. Lonsdale, D. G. Monet, S. Price, P. Seitzer, T. Jarrett, J. D. Kirkpatrick, J. E. Gizis, E. Howard, T. Evans, J. Fowler, L. Fullmer, R. Hurt, R. Light, E. L. Kopan, K. A. Marsh, H. L. McCallon, R. Tam, S. Van Dyk, and S. Wheelock. The two micron all sky survey (2MASS). *The Astronomical Journal*, 131:1163–1183, February 2006. DOI: 10.1086/498708 xxii

John Beaver
December 2019

PART I

Space

CHAPTER 1

Tools for Understanding Space

The universe is big. But it is also very fine-grained, and so there is important detail at both extremely tiny and unimaginably huge scales of size and distance. To fit this into the human mind, we need some tools. One of the most important of such tools is the ability to relax: abandon the idea that it can all make sense at once, and in detail. Instead, we learn how to split it up into pieces that are manageable to the human mind, and we then learn how to relate those piece to each other. In the sections that follow, I describe several methods that are useful for making sense of both the bigness and the tininess, while still keeping track of how it all fits together.

Throughout *The Big Picture* I make use of both *scientific notation* and the *SI system of physical units*. For readers unfamiliar with either of these tools, Appendix A provides details.

1.1   POWERS OF TEN

The diameter of a grape is roughly 10 mm greater than that of a pea; the diameter of an apple is approximately 90 mm greater still.[1] But to say that the diameter of Earth is 6,384,000,000 mm more than that of an apple tells us little. To meaningfully compare sizes that are so disparate, it is not how *much* but rather how many *times* bigger that is important.

The *logarithm* is a mathematical tool that compares numbers by *factors* rather than differences. Its foundation is the *exponential function*. As an example of an exponential function, a sequence of numbers such as 1, 2, 4, 8, 16, 32, … can be written instead as follows: $2^0, 2^1, 2^2, 2^3, 2^4, 2^5, ...$. And so we can describe the sequence with the exponential function $y = 2^x$.

This particular example is an exponential function with a *base* of 2, but other bases may be used; the most common is the *base 10 exponential function*:

$$y = 10^x. \tag{1.1}$$

And so values for x of 1, 2, 3, 4, … simply represent y values of 10, 100, 1000, 10,000, …, or the number's *power of 10*. The inverse of this particular exponential function is called a *common logarithm*, and it is defined such that:

$$\log(10^x) = x. \tag{1.2}$$

And so $\log 10 = 1$, $\log 100 = 2$, $\log 1000 = 3$, …. For numbers smaller than one, logarithms have negative values. For example, $0.01 = 10^{-2}$, and so from Equation (1.2), we must

---

[1]Parts of this section appeared, in a somewhat different form, in Beaver [2018b, Sec. 1.6].

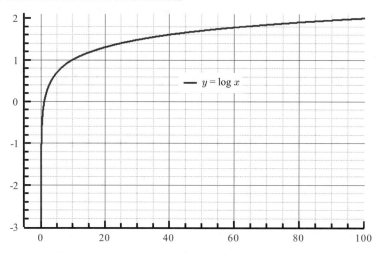

Figure 1.1: A graph of the common logarithm ($\log x$) is defined in terms of powers of 10, and so depicts 0.01 as -2, 0.1 as -1, 10 as 1, and 100 as 2.

have $\log 0.01 = -2$. But although the logarithm of a number can be negative, there is no such thing as the logarithm of a negative number. Furthermore, there is no value, x, for which $10^x = 0$. And so although a logarithm can be zero ($\log 1 = \log 10^0 = 0$), the logarithm of zero is undefined. A graph of the common log function can be seen in Figure 1.1.

Logarithms have interesting and useful mathematical properties, the most important of which are the following:

$$\log(ab) = \log a + \log b \qquad (1.3)$$
$$\log a^b = b \log a. \qquad (1.4)$$

Because of Equation (1.4), logarithms "undo" exponents, turning them into simple factors. And Equation (1.3) shows that logarithms turn multiplication into addition. For this reason, published tables of logarithms were important in pre-computer times, since addition (and subtraction) is easier to perform by hand than multiplication (or division). This led to the old joke that certain types of snakes only reproduce when placed on a wooden table.[2] Apart from that, some species of rodent naturally think in terms of logarithms; see Figure 1.2.

The left side of Figure 1.3 shows the exponential function $y = 10^x$ plotted vs. x, while the right-hand graph shows that same function plotted with a *logarithmic scale* for the y-axis. On a logarithmic scale, each tic mark of the graph represents not an amount, but rather a *factor*. For the example shown here, each tic is 10 times greater than the tic below it. Since $\log(10^x) = x$, this is simply a straight line. And so use of a logarithmic scale on the vertical axis of a graph has the effect of turning an exponential function into a straight line. Notice how the logarithmic

[2]Even adders can multiply on a log table.

Figure 1.2: Of all members of the order *Rodentia*, *Castor canadensis* is most attuned to thinking in terms of powers of ten (Beaver image CC BY-SA 2.0).

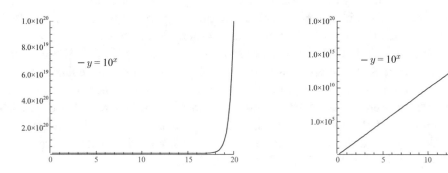

Figure 1.3: The exponential function $y = 10^x$ plotted on an ordinary linear scale (left), and a logarithmic scale (right). The logarithmic scale straightens the exponential function and makes a large range of values more manageable.

scale compresses an enormous range of values to a much smaller scale. Most of the direct graph of the exponential function, as shown on the left, is almost useless; at the scale of the graph, 90% of it is either indistinguishable from zero or nearly vertical.

And so a logarithmic scale is especially useful when trying to compare numbers to each other that encompass an enormous range of values. We represent numbers by their exponents, using a logarithmic scale to compare their *powers of ten*, rather than the numbers themselves.

The *difference* between the powers of ten for two numbers is then by how many *factors* of ten they are apart. Powers of ten are also often referred to as *orders of magnitude*. And so the number 10,000 is three orders of magnitude greater than the number 10:

$$\frac{10^4}{10^1} \;=\; 10^{4-1}. \tag{1.5}$$

There is a famous documentary film called *Powers of Ten* made in 1977 by the influential designers and filmmakers Charles and Ray Eames [Eames and Orear, 1979]; the film is based on a 1957 book called *Cosmic View: the Universe in Forty Jumps* by Kees Boeke [Boeke, 1957]. The view moves outward from a picnic scene, but on a logarithmic scale. Every 10 s the viewer is 10 *times* farther away. Every 10 s the distance across the image is 10 *times* greater than it was 10 s earlier. In this way, 23 orders of magnitude are covered in less than 4 min, bringing the viewer from a picnic basket to the largest-scale structures in the universe. The film then returns to the original scene and works its way inwards to the very fundamental particles that make up matter.

The Eames' use of a logarithmic scale to visually zoom through many orders of magnitude of distance was innovative and an impressive technical feat at the time. This can now be easily accomplished with computers. Google Earth is a good example; its zoom feature works in this way.

Table 1.1 shows the typical size scales—as measured in meters—of some different parts of the universe, expressed with only their powers of ten. And so, for example, the average height of a human is roughly $1.7 \, \text{m}$, and in exponential notation, this is equivalent to 1.7×10^0 m. But in Table 1.1, I have rounded this in order to determine its order of magnitude.

Thus, we can see from Table 1.1 that the Sun, for example, is roughly $10^{9-(-4)} = 10^{13}$ times larger than the thickness of a human hair. And notice that the largest scale listed in the table is an incredible *61 orders of magnitude* greater than the smallest. I have expressed the lengths in meters; if different units were used, the corresponding orders of magnitude would be different as well. But whatever the unit of length, there would still be 61 orders of magnitude between the smallest (the Planck length) and the largest (the particle horizon). We will discuss the meaning of many of the items in Table 1.1 throughout *The Big Picture*.

1.2   MEASURING DISTANCES IN SPACE

1.2.1   USING TIME TO MEASURE DISTANCE

In a vacuum light travels at the always-constant speed of $2.998 \times 10^8 \, \text{m s}^{-1}$. This speed is one of the physical constants of nature, and it is so important that it has its own special symbol, c. We can measure its value, but it is unknown why it has the precise value that it does. An important property of light is that the speed (as measured in a vacuum) is independent of the relative motion of the observer and the source of light. This rather-odd fact is well established by experiment, and it is part of the foundation of Einstein's special and general relativity (see Part IV).

Table 1.1: The size scales of different parts of the universe, and their associated orders of magnitude (powers of ten)

Object	Typical Size (m)	Order of Magnitude
Planck length	1.6×10^{-35}	-35
Atomic nucleus	1×10^{-19}	-19
Atom	1×10^{-10}	-10
Bacterium	1×10^{-6}	-6
Red blood cell	7×10^{-6}	-5
Human hair thickness	1×10^{-4}	-4
Adult human height	1.7	0
Earth	1.3×10^{7}	7
Jupiter	1.4×10^{8}	8
Sun	1.4×10^{9}	9
Earth-sun distance	1.5×10^{11}	11
Sun to α Centauri	4.1×10^{16}	16
Thickness of Milky Way galaxy	2×10^{19}	19
Diameter of Milky Way galaxy	1×10^{21}	21
Size of Local Group	1×10^{23}	23
Distance to Virgo cluster	5.1×10^{23}	24
Universe particle horizon	1.3×10^{26}	26

A speed is a distance per time, and with such a large speed as this, one can get a mental grip on either the distance or time, but not both. And so if we talk about an intuitive interval of time—one second, for example—then the distance light travels is outside our direct intuitive experience. 300,000 km is nearly seven times around the circumference of Earth.[3] If, on the other hand, we make the distance small enough to fit our everyday experience, then the *time* required for light to travel that distance is too tiny to grasp directly.

Because speed is a distance traveled, d, divided by the time, t, to make the trip, we can use c to translate between space and time:

$$c = d/t \tag{1.6}$$
$$d = ct \tag{1.7}$$
$$t = d/c. \tag{1.8}$$

[3]It is also roughly the distance traveled by an automobile in the U.S., after a dozen years of moderately heavy driving.

And so we can use Equation (1.7) to define new units of length. A "light-nanosecond," for example, is the distance light travels in one billionth of a second. It is not difficult to see that this is very nearly 30 cm, or just a bit shy of one foot. For astronomers, the *light year* (ly) is more useful: 1 ly is the distance light travels in one year:

$$1\,\text{ly} = 2.998 \times 10^8\,\text{m/s} \cdot 365.26\,\text{days} \cdot 8.640 \times 10^4\,\text{s/day} \tag{1.9}$$
$$= 9.461 \times 10^{15}\,\text{m}. \tag{1.10}$$

An important example of the relation between c and distance is our very definition of our fundamental unit of length, the *meter*. Because physicists can directly measure time more precisely than length, a meter is defined to be the distance light travels in a particular tiny agreed-upon length of time:

A meter is the distance light travels in 1/299,792,458 of a second.

To precisely measure a distance then, we simply send out a brief pulse of light, let it reflect off some object, and then measure the time required for the pulse to return. For astronomers, this method is limited to nearby rocky bodies in the solar system, close enough and solid enough that we can reflect electromagnetic waves off them. It forms our best distance measurement to the Moon and the planet Venus.

1.2.2   ANGULAR DIAMETER AND PARALLAX

When I was quite young, I once slept on the floor at the house of my cousins. In the middle of the night, I awoke to see, in the dim moonlight streaming through the window, a huge figure, at least 8 feet tall, silhouetted against the wall. It had pointed ears and swayed menacingly from side to side as I lay there in frozen terror.

Then the kitten touched my nose with its own, and it all suddenly shrank to a small and playful ball of fur only inches from me.

It is important to distinguish between how big something *appears* and how big it *is*. We call the apparent size of an object its *angular diameter*, which I will denote with the symbol θ_D (θ is the Greek letter theta). Angular size is measured in angle units, because it is the angle formed when one points toward both sides of the object. Figure 1.4 shows the geometry for a spherical object of radius, R, that is located a distance, d, from the viewer's eye.

It should be clear from Figure 1.4 that if the object is very far away compared to its radius, the angular diameter is small. That is to say, if $d \gg R$, then θ_D is very small.[4] This special case applies to many astronomical objects; they are much farther away than their physical size, and so they appear very tiny in the night-time sky. But also, whenever θ_D is small, we can use a very simple equation—called the *small-angle formula*—to relate the size, distance and angular size. See Equations (1.11)–(1.13) for the three different ways to rearrange the small-angle formula, where I have replaced the radius, R, of the object with its diameter, $D = 2R$. If the object is *not*

[4]I use the symbol $\gg$ to mean "much greater than."

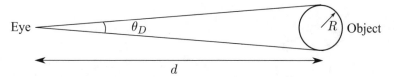

Figure 1.4: The apparent (angular) size of an object depends on both its radius and its distance from the viewer.

small compared to its distance—if the angular size is large rather than tiny—then the simple small-angle formula will produce a significant error, and trigonometry must be used instead:

$$\theta_D = \frac{D}{d} \tag{1.11}$$

$$D = d\theta_D \tag{1.12}$$

$$d = \frac{D}{\theta_D}. \tag{1.13}$$

It is Equations (1.12) and (1.13) that are most important to astronomers, for the simple reason that we can usually measure the angular diameter θ_D directly and easily, and so we don't need to calculate it. The angular size is, after all, simply how big the object appears in the sky, as seen from Earth. Once we have measured the angular size of an astronomical object, we can use Equation (1.12) to calculate its true size—its diameter, D—if we can somehow figure out its distance, d, by some other method. Alternatively, if we already know the object's diameter, then we can use Equation (1.13) to calculate its distance. Astronomers use both of these tricks, depending upon the situation and what information they are able to gather by other methods.

The small angle formula as presented in Equations (1.11)–(1.13) produces correct numerical results only if the angular size is measured in *radians* rather than degrees. The radian measure of an angle is simply the ratio of circular arc swept out by the angle, divided by the radius of the circle. Imagine a slice of pie; the angle of the slice in radians is the length along the edge of the crust divided by the radius of the pie.

A complete circle in radians is then the full circumference of the circle divided by its radius, and that is exactly equal to 2π, since the number π is *defined to be* the circumference of a circle divided by its diameter (which is twice the radius). Thus, there are 2π radians in a complete circle, and so 2π radians = 360°. And so 1 radian is equivalent to $360/2\pi = 57.296°$. Similarly, $1° = 0.01745$ radians. Since a full circle is 2π radians, we have the curious fact that half of a pie is equal to π radians: pi = pie/2.

Angular diameters can be easily measured because the triangle shown in Figure 1.4 is reproduced, at a much smaller scale but with the same angle, in any optical instrument used to record an image of the object [see, for example, Beaver, 2018a, Ch. 8]. As a low-precision example, the left side of Figure 1.5 shows my simple optical device for directly recording an image of the sun (I used it as an eclipse camera). It is simply a hollow tube with a small, simple

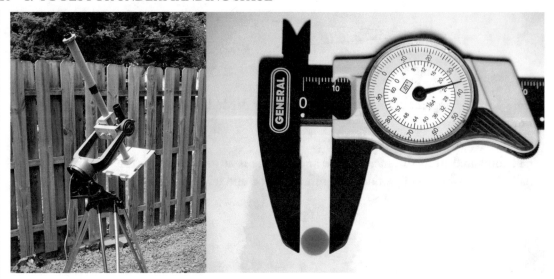

Figure 1.5: A lens at the top of a tube 0.925 m in length (left) formed the 9-mm diameter image of the Sun at right. The small-angle formula allows one to use this data to easily calculate the angular size of the Sun.

lens at one end and a holder for light-sensitive paper at the other. For this particular camera, the image distance is 0.925 m from the lens. The right side of Figure 1.5 shows a direct image of the sun recorded with this camera,[5] with a measured image size of about 9 mm, or 0.009 m. We can thus find the angular size, in radians, of the Sun simply by dividing the image size by the image distance. And so our rough estimate of the angular diameter of the Sun from this crude data is simply $\theta_D = 0.009/0.925 = 0.0097$ radians $= 0.56°$, very close to the actual angular diameter of $0.53°$ on that particular day.

1.2.3   TRIANGULATION, PARALLAX, AND THE ASTRONOMICAL UNIT

We can sometimes use a geometry similar to that of Figure 1.4 in order to directly measure a distance in space. The method of *parallax* is to view the same object from two vantage points that are a known distance apart. And so we form a triangle somewhat like that of Figure 1.4, except that our two vantage points make up the small end of the triangle, rather than its vertex.

The geometry of an important historical example is illustrated in Figure 1.6. In 1761 and 1769 the planet Venus passed directly between Earth and the Sun; such an event is called a *transit*. During this rare occurrence, astronomers at different locations on Earth measured the time required for the silhouette of Venus to cross the Sun's disk [North, 1995, p. 253]. But

---

[5]It is a negative photographic process, so the Sun appears as a dark disk on a white background.

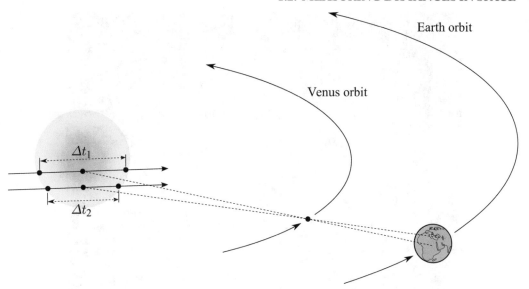

Figure 1.6: Parallax was used to measure the distance between Earth and Venus, by making observations from two different locations on Earth of a transit of Venus across the Sun's disk. (Illustration by Duckysmokton, Ilia, Vermeer, CC BY-SA 3.0.)

because of the slightly different vantage points, different observer's saw the planet transit at different solar latitudes, and so require different amounts of time.

It is possible to relate those time differences, given the known angular diameter of the Sun, to the small angle (much exaggerated in the illustration) of the triangle that is formed by the geometry. The distance between the two observation points on Earth, then, is like the right side of the triangle in Figure 1.4. We can thus use the small-angle formula to calculate the distance, d, to Venus:

$$d = \theta B, \tag{1.14}$$

where B is the *baseline*—the physical distance between the two observation points.

Using parallax to triangulate the true distance to Venus provided the scale factor for the entire solar system. Thus, these eighteenth-century studies of transits of Venus allowed for the first real measurements of the Earth–Sun distance—the *astronomical unit* (AU).

Another transit of Venus occurred in June of 2012; see Figure 1.7 for my photograph of one stage of the transit. The dark circular disk is the silhouette of Venus, while the smaller not completely dark spots are sunspots. The true size of Venus is not much bigger than the largest of these sunspots. It appears much larger that that because it is much closer than the Sun.

Figure 1.7: My photograph of the June, 2012 transit of Venus. Contrast the perfectly round silhouette of Venus with the irregular and less-dark sunspots. Since Venus is much closer than the Sun, it appears much larger than its true size, which is similar to that of the largest sunspots shown here.

1.2.4   STELLAR PARALLAX

We can use the method of parallax to measure the distances to nearby stars. But a far larger baseline, B, is needed because the distance, d, is so much greater than for the case discussed in Section 1.2.3. Instead of observing from different locations on the surface of Earth, we allow our planet's orbital motion about the Sun to carry us to different parts of Earth's *orbit*. Thus, our baseline over the course of a year is a full 2 AU; Figure 1.8 shows the basic geometry.

There are many circumstances where angles measured in astronomy are extremely small—a minuscule fraction of a degree. Such is the case for the parallax angle measured for even the closest of stars. Since the radian is already a large angle (over $57°$), it is inconvenient for everyday descriptions of these tiny angles. And so astronomers commonly use other units of angle instead, only converting to radians when a calculation is needed. The most common units for tiny angles are the *arcminute* (or *minute of arc*) and the *arcsecond* (or *second of arc*). These units have nothing *per se* to do with time; they are simply fractions of a degree: there are 60 arcminutes per degree, and 60 arcseconds per arcminute. They are commonly denoted with single and double quotes,

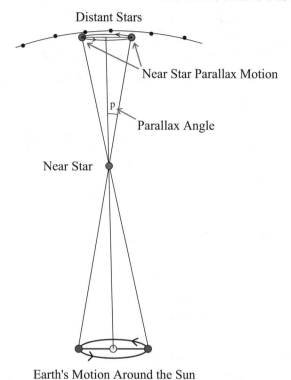

Figure 1.8: *Stellar parallax:* a relatively nearby star appears in slightly different directions as seen (against the backdrop of very distant stars) from two vantage points on opposite sides of Earth's orbit about the Sun. (Graphic public domain)

just like minutes and seconds of time. And so we have the following:

$$1 \text{ arcminute} = 1' = 0.01667° \tag{1.15}$$

$$1 \text{ arcsecond} = 1'' = 0.01667' = 0.0002778° \tag{1.16}$$

$$1 \text{ radian} = 57.296° \tag{1.17}$$

$$1° = 3600'' \tag{1.18}$$

$$1 \text{ radian} = 206,265''. \tag{1.19}$$

Astronomers take a practical approach to measurements of stellar parallax, taking advantage of the following.

- The baseline is always the same for measurements of stellar parallax—the diameter of Earth's orbit.

- The stellar parallax angle is always very tiny, and so it is measured in arcseconds. Even for the nearest star, this angle is *less* than one arcsecond.

- We define the stellar parallax angle in terms of the right triangle made by bisecting the isosceles parallax triangle, as shown in Figure 1.8. And so relative to this angle, the baseline is the *radius* of Earth's orbit—that is, the astronomical unit itself.

And so we can define a new unit of distance, the *parsec*, as follows:

A parsec (pc) is the distance an object (such as a star) would need to be in order to have a stellar parallax of exactly 1 arcsecond.

Notice that the word "parsec" is simply an invented contraction of the words "parallax" and "second." Since 1 radian = 206, 265″, it is easy to see that the parsec is equivalent to 206,265 AU. This is a huge distance, approximately equal to 3.26 ly.

Given this definition of the parsec, we can see from Equation (1.11) that the following very simple relation must hold between the distance, d, to a star as measured in parsecs and its parallax angle, p, in arcseconds:

$$d(\text{parsecs}) = \frac{1}{p(\text{arcseconds})}. \tag{1.20}$$

Proxima Centauri is the nearest star to us (other than the Sun of course), and with a measured parallax of 0.7686″, it has a distance of 1.301 pc.

1.3 SCALING AND SCALE MODELS

Most of the distances and sizes in the universe are either too tiny or too large to experience directly. But even when the numbers themselves are far outside the range of our direct experience, we can still describe the relationships between those numbers. We may then form a *scale model* that preserves the size and distance relationships, while scaling everything together to sizes and distances that are within the easy reach of human perception.

We will use this technique in Chapter 2 to illustrate the relationship between the sizes and distances of bodies within the solar system. Let us here, as an introduction, set up a simple numerical example for a scale model of the three solar-system bodies most important to humans—Earth, Moon, and Sun.[6]

Earth, Moon, and Sun are all (approximately) spheres, each with its own radius. The Moon orbits Earth at a particular (average) distance, while much farther away the Earth–Moon pair orbits the Sun. In what follows, I use the standard astrological symbols for Earth ($\oplus$), Moon ($\mathbb{C}$), and Sun ($\odot$) as subscripts to denote these bodies, placed on R for radius or d for distance from Earth. And so, for example, $R_\oplus$ means "radius of Earth." In SI units of meters, we have the following data:

$$R_\oplus = 6.380 \times 10^6 \, \text{m}$$
$$R_\mathbb{C} = 1.738 \times 10^6 \, \text{m} \quad d_\mathbb{C} = 3.84 \times 10^8 \, \text{m}$$
$$R_\odot = 6.963 \times 10^8 \, \text{m} \quad d_\odot = 1.496 \times 10^{11} \, \text{m} .$$

[6]Do not look directly at the Sun!

I have put these numbers all in the same units of meters, and this allows them to be directly compared to each other. As described in Section 1.1, we can learn a lot simply by comparing the exponents in the scientific notation. Only a glance is needed to see that Earth and the Moon have sizes that are within a factor of 10 of each other, but the Sun's radius is two powers of ten (a factor of 100) larger. Also notice that the *distance* to the Moon has the same power of ten as the *radius* of the Sun, but the distance to the Sun is three orders of magnitude greater than the distance to the Moon.

To make a scale model out of this data, we multiply all of the numbers by the same *scale factor*. Of crucial importance is that all of the numbers are *lengths* (as opposed to, for example, volumes), and that they are all expressed in the same units, before multiplying by the scale factor. The trick is to choose a scale factor such that all of the scaled lengths are then of a size that is within the range of direct human experience. If these sizes can be related to familiar objects, so much the better.

For this particular example, the numbers are all too *large* for our experience, and so we need to make them smaller in our scale model. Thus, we need a scale factor that is a tiny fraction, much less than unity. Equivalently, we could choose a large scale *divisor* instead of a factor, dividing all of our numbers by the same large number. But whether we multiply by a tiny number or divide by a large one, the goal is to put all of the numbers on a human scale.

We have two rather obvious choices as a starting guess for an appropriate scale factor. On the one hand, we can try to scale down the smallest-size object (the radius of the Moon in this example) to the smallest size we can directly experience—a grain of salt, for example. Or alternatively, we can choose a scale factor such that the *largest* distance (the distance to the Sun in this example) is now just barely small enough to experience directly and easily.

Let us make both of these choices and see what we get. And so let us choose on the one hand a scale factor that puts the Moon down to the size of a grain of salt, scaling all of the other numbers by that same factor. And for comparison, let us choose a different scale factor that puts the Earth–Sun distance at, say, 100 m—roughly the length of a soccer pitch or American football field, a familiar distance that is large, but that one can easily see or walk across.

It is simple to determine the first scale factor example, which I will represent as s_1. We want a value for s_1 such that when it is multiplied by the diameter of the Moon (twice the radius $= 2R_{\mathbb{C}}$), we get a size equivalent to the width of a grain of salt (d_{salt}):

$$2R_{\mathbb{C}}s_1 = d_{\text{salt}} \tag{1.21}$$

$$s_1 = \frac{d_{\text{salt}}}{2R}. \tag{1.22}$$

If we choose the width of a typical grain of salt to be one third of a millimeter (3.33×10^{-4} m), then we have:

$$s_1 = \frac{3.33 \times 10^{-4}\ \text{m}}{2 \times 1.738 \times 10^6\ \text{m}} = 9.58 \times 10^{-11}. \tag{1.23}$$

Table 1.2: Two examples of scale models for the Earth–Moon–Sun system, using two different scale factors

	Real Value (m)	× s_1 (m)	× s_2 (m)
$R_\oplus$	6.380×10^6	6.11×10^{-4}	4.26×10^{-3}
$R_{\mathbb{C}}$	1.738×10^6	1.67×10^{-4}	1.16×10^{-3}
$R_\odot$	6.963×10^8	6.67×10^{-2}	0.465
$d_{\mathbb{C}}$	3.84×10^8	3.68×10^{-2}	0.257
$d_\odot$	1.496×10^{11}	14.3	100

Notice that the units—meters in this case—cancel, leaving a dimensionless number for the scale factor.

We can do the same for our second attempt at a scale factor, s_2, using the largest distance ($d_\odot$) scaled to 100 m:

$$d_\odot s_2 = 100\,\text{m} \tag{1.24}$$

$$s_2 = \frac{100\,\text{m}}{d_\odot} \tag{1.25}$$

$$s_2 = \frac{100\,\cancel{\text{m}}}{1.496 \times 10^{11}\,\cancel{\text{m}}} \tag{1.26}$$

$$s_2 = 6.68 \times 10^{-10}. \tag{1.27}$$

We can now multiply these two scale factors by our data to see what we get. The results can be seen in Table 1.2. Notice that the first scale factor puts the radius of the Moon at half the width of a grain of salt, and the second scale factor puts the distance to the Sun at 100 m. The second scale model is bigger than the first, because our second scale factor was larger.

And so which of the two scale models is the most useful? The first puts the Earth–Sun distance at an easily-manageable 14.3 m—a distance of about 20 steps (assuming a typical step length of 0.7 m)—and the Moon is nicely placed at 3.68 cm from Earth. The Sun is a little over 13 cm, about the size of a large grapefruit. By design, our scaled Moon is the size of a grain of salt. Earth, then would be about 1.2 mm across—still very tiny. For an appropriately scaled Earth for this model, one could purchase a size 16/0 glass seed bead, which is about this size.[7] And so our scale model of the Earth–Moon–Sun system is a tiny glass bead for Earth, a grain-of-salt Moon 3.68 cm away, and a grapefruit Sun some 20 steps distant.

For our second scale model, the Moon and Earth are larger and more manageable in size— 2.3 mm across for the Moon and 8.5 mm for Earth. It would be a rather simple and enjoyable task to wander around with a millimeter scale ruler, looking for suitable Earths and Moons

---

[7]They are, however, only sold in packs of many hundreds, and that is a lot of extra Earths to give to friends as birthday presents.

for this scale model. Finding a nearly 1-m diameter ball for the Sun might be somewhat more difficult. And clearly this model would need more physical space to set up; the scaled Moon is about 10 in from Earth, but the Sun is a full 100 m distant.

This simple example of a scale model is an easy one because the full range of sizes and distances is rather limited. But what if one wanted to include not only Earth, Moon, and Sun, but also Jupiter, Saturn, Uranus, and Neptune? Or what if one were to draw appropriately-sized craters onto our scaled Moon, or include the star Alpha Centauri in our model? This is where the hard choices come to play, because the scale model can easily extend beyond the range of our direct human experience, thus missing the point of making a scale model in the first place. Clearly, a powerful microscope would be needed to see scaled-down craters on our grain-of-salt Moon. And even the nearest star is very far away indeed, even when scaled down so much that the Moon is, by comparison, only a grain of salt.

One nice thing about a scale model is that *angular sizes are preserved*. Consider the first scale model described in Table 1.2; a grapefruit-sized Sun is placed 14.3 m from Earth. We could imagine shrinking ourselves down onto this tiny Earth, and from that vantage point looking at the grapefruit Sun.[8] How big would our scale-model Sun appear? That is to say, what is the angular size of a 13.3-cm diameter grapefruit, when viewed from a distance of 14.3 m? We calculate the answer with the small angle formula:

$$\theta_D = \frac{0.133\,\text{m}}{14.3\,\text{m}} \tag{1.28}$$

$$\theta_D = 9.3 \times 10^{-3}\,\text{radians} \tag{1.29}$$

$$\theta_D = 0.53°. \tag{1.30}$$

And so we see that when viewed from the scale-model Earth, the grapefruit sun appears exactly the same size as does the real Sun in the real sky. This result holds in general for scale models. Since angular sizes are ratios of lengths, the angular sizes of objects in a scale model are the same as in real life.

1.3.1   SCALING RATIOS

Any particular scale model is only useful to portray a limited chunk of the universe. We can, however, use *different* scale models for different parts of the universe, and find ways to put them together so as to compare them to each other. A starting point for this task is to calculate simple comparisons, in the form of numerical ratios, of chosen pairs of sizes and distances. These ratios are the same whether one uses the physical numbers or their scale-model versions. And so for example, one can easily calculate from the numbers in Table 1.2 that Earth has a radius 3.67 times greater than that of the Moon. This is true in real life, but it is also true for any (correctly made) scale model of the Earth–Moon system.

---

[8]It is generally considered safe to look at a grapefruit without eye protection.

Table 1.3: Scaling ratios for sizes and distances of Earth, Moon, and Sun

$R_\oplus$	6.380×10^6 m	$R_\oplus/R_\mathbb{C}$	3.67
$R_\mathbb{C}$	1.738×10^6 m	$R_\odot/R_\oplus$	109
$R_\odot$	6.963×10^8 m	$d_{\mathbb{C}\oplus}/R$	60.2
$d_\mathbb{C}$	3.84×10^8 m	$d_\odot/R_\odot$	215
$d_\odot$	1.496×10^{11} m	$d_\odot/R_\oplus$	23,400

Table 1.3 shows some examples of these *scaling ratios*. Note that the ratios are (and should be) dimensionless, and this requires that both lengths be expressed in the same units before dividing one by the other.

Scaling ratios show us important relationships at a glance. The Sun has 109 times the diameter of Earth. The Moon is at a distance of 60 Earth-radii (30 Earths side-by-side). The distance from Earth to Sun is roughly the same as 100 suns—or 10,000 Earths—placed side-by-side.

1.4   SURFACE AREA, VOLUME, MASS, AND DENSITY

An official NBA basketball should have a circumference of 29.5 inches, or 74.9 cm, for a diameter of $74.9/\pi = 23.9$ cm. An official ITF tennis ball, on the other hand, should have a diameter between 6.54 and 6.86 cm. This means a basketball has $23.9/6.54 = 3.65$ times the diameter of a smaller-sized tennis ball. By an odd coincidence, this is almost exactly the size of the Moon as compared to Earth; the diameter of Earth is 3.67 times the diameter of the Moon (see Table 1.3). Simply over-inflate the basketball slightly, and the comparison to the Earth and Moon could be made exact.

Place a basketball next to a tennis ball (see Figure 1.9) and one could be forgiven for thinking the basketball is considerably more than 3.7 times bigger than the tennis ball. The reason is that when we visually compare spherical objects to each other we seem to do so more in terms of their *surface areas* than their diameters [see, for example, Jansen and Hornbak, 2016]. This suggests we must be careful when we say, for example, that one object is 3.68 times "bigger" than another. Do we mean its diameter, surface area, or volume?

For spherical objects, the relations between diameter (or radius, $R = $ diameter/2), surface area, A, and volume, V, are simple:

$$A = 4\pi R^2 \tag{1.31}$$
$$V = {}^4\!/_3\pi R^3. \tag{1.32}$$

Figure 1.9: The diameter of a tennis ball as compared to a basketball is nearly the same ratio as that of the Moon compared to Earth.

If we mean to only *compare* these quantities for two different objects, it is simpler still:

$$\frac{A_1}{A_2} = \left(\frac{R_1}{R_2}\right)^2 \tag{1.33}$$

$$\frac{V_1}{V_2} = \left(\frac{R_1}{R_2}\right)^3. \tag{1.34}$$

Although Equations (1.31) and (1.32) are valid only for spheres, Equations (1.33) and (1.34) are correct whenever one compares *any* two objects, so long as they both have the same shape. Since surface area scales with radius squared, its SI units are not meters (m), but rather *square meters* (m^2). And it is also clear that volume must be measured not in meters, but rather in *cubic meters* (m^3).

To say that the radius (or diameter) of Earth is 3.68 times the radius of the Moon is to say that Earth's surface area is $3.68^2 = 13.5$ times as great, while its volume is $3.68^3 = 49.8$ times greater. And so although a basketball has only 3.65 times the diameter of a tennis ball, it has over 13 times the surface area, and it is this larger number that is more connected to our human perception when we visually compare the "sizes" of objects.

The surface area of a spherical astronomical body has an important physical significance; it is related to the (perhaps invisible) radiation the body emits. But the *volume* is also important; it tells us something about the total amount of stuff the body is made of. The amount of water needed to fill a hollow spherical container, for example, is directly proportional to the container's volume rather than its surface area or radius.

But there is another important physical quantity that is related to (but not the same as) volume—the object's *mass*. We will consider mass often throughout *The Big Picture*; it is a subtle and complex subject. But as a start, let us consider these two different and very-crude definitions of mass.

1. On the surface of Earth, an object with more mass *weighs* more than an object with less mass.

2. An object with greater mass causes your foot to hurt more when you kick it, compared to kicking an object with less mass.

This first definition is clearly about gravity, and in this crude and simple form it is not of much help; how does one "weigh" the Moon? The second definition is called *inertia*, and it seems to be distinct from gravity. Bring a cathedral with you to a blank region of interstellar space—so gravity is an insignificant factor—and then kick it. Your foot will hurt just as much as if you had kicked that same cathedral here on Earth. We will make both of these definitions of mass more precise as we go along. And eventually we shall see that these two seemingly distinct ways to look at mass are not so different as they seem.

Clearly, mass has something to do with the total amount of "stuff" an object is made of. And so there is a connection between mass and volume. If one lead balloon[9] has twice the volume as another, it also has twice the mass. But of course it is not *only* about volume; a lead balloon would have far more mass than an ordinary air-filled balloon of the same volume. And so there is some property of the *type of material* itself—lead vs. air, for example—that is also connected to the mass of an object. That property is *density*,[10] and we define its average value for some object in this way:

$$\bar{\rho} = \frac{m}{V},\tag{1.35}$$

where m is the object's mass and V is its volume, and we by tradition employ the Greek letter ρ (rho) as the symbol for density, and $\bar{\rho}$ for its average value.

Equation (1.35) is only the *average* density of the object (we place a bar over a variable to represent its average value); it may be made of different materials, with some parts of higher density and others of lower density. But one can take any arbitrary piece of an object and use Equation (1.35) to calculate the average density of that part. If one chooses a tiny-enough part, such that the material throughout is all the same, then the calculation represents the density, ρ, of the material *at that particular point* in the object.

The average density of an object provides important clues regarding the type of material it is made of. On the other hand, if the composition of the object (and thus its density) can be guessed, we can then measure its volume and rearrange Equation (1.35) to calculate its mass:

$$m = \bar{\rho}V.\tag{1.36}$$

---

[9]I should make some sort of humorous comment at this point, but I fear it would go over like ....

[10]My jokester father has been known to say that an object of high density is "heavy for its weight."

In practice, even if an astronomical object (the Sun for example) is made of the same material throughout, Equation (1.36) is often too simplistic because most materials—gases in particular—are compressible. The gas deep within the Sun, for example, has a much higher density than the gas farther from the center even though it is made of the same mix of chemical elements.

1.5   REFERENCES

John Beaver. *The Physics and Art of Photography, Volume 1: Geometry and the Nature of Light*. IOP Publishing, 2018a. DOI: 10.1088/2053-2571/aae1b6 9

John Beaver. *The Physics and Art of Photography, Volume 3: Detectors and the Meaning of Digital*. IOP Publishing, 2018b. DOI: 10.1088/2053-2571/aaf0ae 3

Kees Boeke. *Cosmic View: The Universe in 40 Jumps*. John Day Company, New York, 1957. 6

Charles Eames and Jay Orear. Powers of ten-1978 (film). *American Journal of Physics*, 47(3):297–297, March 1979. DOI: 10.1119/1.11851 6

Yvonne Jansen and Kasper Hornbak. A psychophysical investigation of size as a physical variable. *IEEE Transactions on Visualization and Computer Graphics*, 22(1):479–488, January 2016. DOI: 10.1109/tvcg.2015.2467951 18

John North. *The Norton History of Astronomy and Cosmology*, W.W. Norton, New York, 1995. 10

CHAPTER 2

Looking Outward

2.1 EARTH, MOON, AND SUN

2.1.1 THE SHAPE OF THE EARTH

Anyone can look out the window and see that Earth is flat. I say this with my tongue only partly in cheek, because there is a sense in which it is true. The Earth is very nearly *locally* flat; it is very difficult to come up with an experiment to prove Earth is a sphere, if one uses *only* evidence from within the small confines of a laboratory. Peer out the window of the laboratory for a prolonged period of time, on the other hand, and other evidence comes into play. The stars appear to take a daily (*diurnal*) trip around us, and the illusion is that they are all attached to a vast *celestial sphere*. The Sun and Moon seem to do the same, but not quite; they have their own much-slower apparent motions superimposed upon the diurnal motion of the celestial sphere.

What is the best explanation for these observations? The devil is in the details of the word "best," and to pick that apart is to explore much of the history of the birth of modern science. But one obvious idea has been articulated since at least ancient Greece: Earth rotates on an axis once per day, and carries us along with it. Of course, that fact alone does not preclude the idea that it is a *flat* earth that rotates. But it is easy to bring in other kinds of non-local data that does not square with a flat-Earth hypothesis.

Consider what happens when one looks at the sky from *other* locations on Earth. The left side of Figure 2.1 shows the position of the stars in the sky as seen at midnight, when facing South from Appleton, Wisconsin on New Year's Day at 10:30 pm. The familiar constellation Orion is high in the sky. The right side of Figure 2.1 shows a photograph I took of those same stars, also on a night in early January, but from central Chile, in the southern hemisphere. From the point of view of a Cheesehead in Chile, Orion is upside down.

We can add even more evidence by traveling to other laboratories, in other parts of Earth, and making similar observations and comparing them. These observations, taken together, show that the following is true.

> The *zenith*—the point directly overhead—points toward different directions in space as seen from different parts of Earth.

This basic observation is easy to make in the modern world. Simply find a friend thousands of miles away and use Internet instant messaging to ask them to precisely describe the position of the Sun, Moon or stars right now. These observations are perfectly consistent with a (roughly) spherical Earth, surrounded by a *very distant* Sun, Moon, and stars. And the zenith—"up"—

Figure 2.1: The familiar constellation of Orion. **Left:** The view from Appleton, Wisconsin (USA), at roughly 45° North latitude (as mapped by the software package Stellarium, available from stellarium.org). **Right:** As photographed from near Casablanca, Chile, at about 33° South latitude (photograph by the author).

means *away from the center of Earth*. There is no way to reconcile these observations with a flat Earth without playing hard and fast with easily verifiable facts.

But why *do* we all experience our local zenith to be away from the center of Earth? The answer is, of course, *gravity*—a topic we take up in detail in Part IV. But let us ask a related question. Is it possible to determine that Earth is a sphere without looking at the sky? Could one move around to different parts of Earth, making only *local* measurements as one moves, and so prove Earth is a sphere? Let us imagine making the following three-legged trip.

1. Start at the North Pole, and travel due south along the 78° 30′ W longitude line to the equatorial city of Calacali, Ecuador.

2. From Calacali, turn 90° to the left (due east) and travel in a straight line along the equator until reaching longitude 11° 30′ E, in central Gabon.

3. From central Gabon, again turn 90° to the left (now heading due north) and travel back along that longitude line to the North Pole.

This imaginary journey forms a triangle—three points connected by straight lines; see Figure 2.2. Our triangle seems to have curved lines, and in three dimensions they *are* curved. But if we constrain ourselves to the two-dimensional surface of Earth, there is an important sense in which the lines of our triangle are straight; *they are the shortest possible paths between their end points*. This is, in fact, the mathematical definition of a straight line. We could—in our imagination anyway—tediously use trial-and-error to find these shortest paths between the

Figure 2.2: A triangle on the surface of Earth, plotted with Google Earth. The three lines are *straight* on the two-dimensional surface; they are the shortest possible paths between their endpoints. (Images copyright 2018 by Google, U.S. Department of State Geographer, image Landsat/Copernicus, Data SIO, NOAA, U.S. Navy, NGA, GEBCO.)

three points of our triangle, even with no knowledge of longitude and latitude or that Earth is a sphere.

Our imaginary journey would thus form what seems to be an ordinary, albeit huge, triangle. It is three points on the surface of Earth, all connected by straight lines—the shortest possible paths between the points. But this would be an odd sort of triangle indeed. One of the first rules of plane geometry is that the inside angles of any triangle must add to 180°. But see Figure 2.3; the interior angles of our giant Earth triangle add instead to 270°. Also, notice from Figure 2.3 that when we zoom in to a small region of our triangle, the lines do indeed appear straight, not curved. And so from the *local* point of view of making our triangle, nothing would seem to be amiss. It is only when we compare the results from different locations—measuring the angles at each vertex and adding the results—that we find something unusual.

Faced with such experimental evidence, we would be left with two possible explanations for our observations. Either the surface of Earth is not flat—it has *curvature* into the third

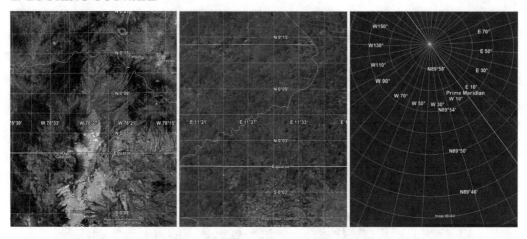

Figure 2.3: All three vertexes of the triangle shown in Figure 2.2 are right angles, meaning the interior angles add to 270°, rather than the 180° the angles would add to if the triangle were on a flat surface. Over a small enough region, the curved surface of Earth is indistinguishable from flat. (Images copyright 2018 by Google, U.S. Department of State Geographer, image Landsat/Copernicus, Data SIO, NOAA, U.S. Navy, NGA, GEBCO.)

dimension, consistent with the surface of a sphere. Or else the rules of plane geometry we learned in high school are somehow incorrect, and it is only *small* triangles that obey the 180° rule.

We know of course that the first explanation is the correct one for our giant Earth triangle; Earth is a sphere. But it turns out that there *are* other geometries besides the plane geometry familiar from high school, and some of these *non-Euclidean geometries* do indeed allow the interior angles of a triangle to add to more than 180°, *even in three dimensions*. This is an important part of Einstein's explanation for gravitation, and we take it up in more detail in Part IV of *The Big Picture*.

2.1.2   THE ROTATION OF EARTH

Figure 2.4 shows two time-exposure photographs of the stars; the shutter of the camera was left open for roughly 10 min in each case. The stars show up as streaks, demonstrating the apparent diurnal motion of the celestial sphere. The picture on the left was taken with the camera pointing north, from Glacier National Park in Montana, while the right-hand picture was taken from central Chile, in the Southern hemisphere, with the camera facing southward.

Both images show the stars seeming to swirl around a particular point on the celestial sphere. For the northern example, a star (called Polaris) happens to be at the pivot point; the southern hemisphere example on the other hand is a relatively blank region of the sky.

One *could* explain these observations by assuming the stars are attached to a very large sphere that is vastly larger than the diameter of Earth. And we could imagine that this *celestial*

Figure 2.4: Time exposures of stars, showing Earth's rotation; the camera shutter was left open for approximately 10 min. **Left:** from Glacier National Park in Montana, in the Northern Hemisphere. **Right:** from central Chile in the Southern Hemisphere. The seeming pivot points are directly above Earth's north pole (left) and south pole (right). Photographs by the author.

sphere rotates around Earth once per day. The seeming pivot points in our picture would then represent the actual southern and northern points of the rotation axis of this celestial sphere.

We know of course that there is a much simpler explanation; the stars are relatively fixed in space, and it is the spherical Earth that rotates on its axis. It is gravity that determines what is up and what is down; up is simply away from the center of Earth. And so as Earth turns on its axis from west toward the east carrying us along with it, the stars appear to move in the opposite direction, like a slower car seeming to go backward as one passes it on the highway. And the northern axis of Earth happens by chance to point very nearly in the direction of the naked-eye star Polaris (while the southern axis points toward nothing in particular).

But how do we know it is this second explanation, rather than the first, that is correct? To the modern mind, the first explanation seems absurd on the face of it. We humans seem to have a natural tendency to prefer the simpler explanation, whenever we are confronted with two explanations that agree equally well with known facts. This is sometimes called *Occam's razor*, and it is an important part of the scientific process.

But a rotating Earth—instead of a rotating celestial sphere—is the simpler explanation only from the point of view of much else that we now know. To the ancient mind, the notion of a rotating Earth raised troubling questions. The ancient Greeks, for example, had measured the size of Earth (see Section 2.1.4), and so it was easy to see that a point on the equator would be carried along by the rotating Earth at nearly 1700 km/hr (over 1000 mi/hr). Why don't we fly off? It is only in the light of workable theories of gravitation and motion that this quite-reasonable concern can be properly addressed, and these theories were not developed in full until the late 17th century (primarily by Isaac Newton).

After Newton's synthesis of motion and gravity (discussed in detail in Part IV), there was no longer any reason to prefer a rotating celestial sphere over a rotating Earth. And so nearly all scientists since that time assumed it is Earth that rotates, even though more than 150 years would pass before there was any direct experimental evidence for a rotating Earth.[1] The results of the *Foucault Pendulum* experiment, publicly demonstrated in 1851 by Léon Foucault, are exactly as predicted for a rotating Earth.

I call attention to this history as an example of an important fact about science. We believe a scientific theory to be true not necessarily because it has been "proven" in some logical manner—such a mathematical proof is usually possible only for rather trivial details. Rather, we believe a theory when, given all else that we know, it seems *un*reasonable to *dis*believe it (see, for example, Beaver [2018], Ch. 1 and references therein).

2.1.3   THE SPHERICAL EARTH: AN EXAMPLE

Figure 2.5 illustrates a personal example that combines both the rotation of Earth and the non-Euclidean geometry of its two-dimensional surface. On a flight from Chicago, Illinois to Edinburgh, Scotland, the rising full Moon was directly out the window next to my seat. The full Moon is directly opposite the Sun in the sky, and so it rises just at sunset (left-hand image). Notice the dark band just above the horizon; it is, essentially, the shadow of Earth. The full Moon in the picture is very near the edge of that shadow, demonstrating that it is very nearly opposite the Sun.

The second image was taken hours later, after waking briefly during the overnight flight. But the Moon was still located directly out my window, and this is an odd thing. For not only was Earth turning, carrying its surface, atmosphere and the Boeing 757 with it, but the aircraft was flying at nearly $900 \, \text{km} \, \text{s}^{-1}$ in roughly the same direction relative to Earth's surface. And so over the course of a few hours, one would think the roughly Southeast-facing wing of the aircraft would be pointing far to the left of the Moon, leaving it well behind in my view out the window.

If the aircraft had been flying merely due East, along a latitude line, that would indeed have been the case. Edinburgh is at a considerably higher latitude than Chicago, so of course the flight initially traveled generally northeast. But the heading of the plane changed drastically over the course of the flight. You can see something close to the actual path of the plane in Figure 2.5, where I used the ruler tool in Google Earth. The flight began with a heading of about 47° North of East, but by the time the aircraft was over the Labrador sea, it was headed only 15° North of East. And it crossed 30.5° longitude heading due East. Flying over Scotland toward Edinburgh, the plane was headed 22° *South* of East.

---

[1] There were important earlier hints as well, in the late 16th and early 17th centuries, from Galileo's groundbreaking studies of physics, the observations of planetary motion by Tycho Brahe, and the investigations of Johannes Kepler.

Figure 2.5: While flying across the Atlantic Ocean, the Moon was visible out my window for hours, even though both the surface of the rotating Earth and the high-speed motion of the aircraft progressed Eastward. (Photographs by the author.)

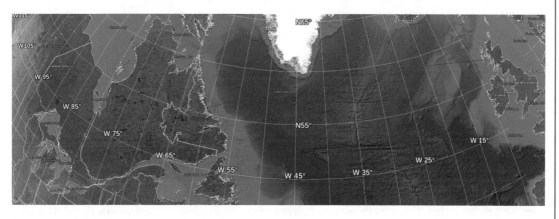

Figure 2.6: The shortest path along the surface of Earth—a geodesic—connecting Chicago and Edinburgh extends to latitudes more northerly than either city. The path looks curved compared to the grid of longitude and latitude. (Image copyright 2018 by Google, U.S. Department of State Geographer, image Landsat/Copernicus, Data SIO, NOAA, U.S. Navy, NGA, GEBCO.)

And so, relative to our grid of longitude and latitude, the airplane followed a curved path, and its heading slowly rotated as it flew, partly counteracting the effect of Earth's rotation regarding the view of the Moon out my window. But this seemingly curved path is in an important sense not curved at all; it is a *geodesic*—the shortest path along the curved surface of Earth, as discussed in Section 2.1.1. Stretch a string on a globe between the two cities and it will naturally follow such a shortest path, and to use the least amount of fuel, flights over long distances also travel roughly along geodesics.[2]

2.1.4   THE SIZE OF EARTH

And so let us assume that Earth is a sphere, and that "up" is simply the direction away from its center. If we also assume that the Sun, Moon, and stars are very far away, compared to the diameter of our spherical Earth, it is straightforward to use observations of these heavenly bodies to measure the size of Earth. We need only compare the directions toward the *same* heavenly body, as seen from two *different points on Earth*.

The first known demonstration of this was described in ancient Alexandria by Eratosthenes, around 240 BC. The angle with the vertical of the noonday Sun was measured on the same date from two cities on a north-south line, simply by noting the length of a shadow cast by a vertical stick of a certain length. The difference in these two angles, given as a fraction of a circle, is then simply the fraction of the circumference of Earth that one travels in the distance between them. See Figure 2.7.

And so our Earth is a sphere with a circumference of about 40,000 km and a radius of 6400 km. By comparison, automobiles are driven per year on average about 12,500 km in the U.K., and about 16,000 km in the U.S.

2.2   THE SOLAR SYSTEM

2.2.1   A SCALE MODEL OF THE SOLAR SYSTEM

Is it possible to simply look up at the night time sky, with no means of directly measuring distances, and still be able to determine, for example, how far away is Jupiter? The answer is yes, so long as we confine ourselves to *relative* distances. But even then it is not so simple; we must have *some* theory as to how the planets are arranged in space.

For the ancient Greeks, the question "how are the planets arranged in space," was perhaps not meaningful. We now think of a "planet" as a corporal body that moves through three-dimensional space, according to physical laws that act upon it. But to the ancients, the planets were points of light in the sky that had their own apparent motions separate from those of the stars. And it was not at all clear that whatever laws may apply on Earth, that those same laws applied to this celestial realm. In antiquity, Earth was not a planet.

This all changed, and it changed gradually. But a seminal point was what is know as the *Copernican Revolution* of the 16th–17th centuries. The apparent motions against the background of stars of the naked-eye planets—Mercury, Venus, Mars, Jupiter, and Saturn—are very complex. On any given day they seem to move along with the celestial sphere in the same manner as the stars, Moon, and Sun. But upon closer inspection, over a longer period of time, it is apparent that they have their own much-slower motions. The very name *planets* means "wanderer."

---

[2]There are of course other considerations that affect the precise flight paths taken.

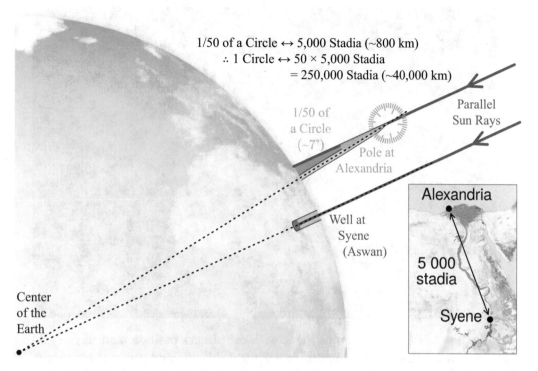

1/50 of a Circle ↔ 5,000 Stadia (~800 km)
∴ 1 Circle ↔ 50 × 5,000 Stadia
= 250,000 Stadia (~40,000 km)

1/50 of
a Circle
(~7°)

Parallel
Sun Rays

Pole at
Alexandria

Alexandria

Well at
Syene
(Aswan)

5 000
stadia

Center
of the
Earth

Syene

Figure 2.7: The differing angle of noonday sunlight between two points on Earth is easily measured. If the physical distance between the two observation points is known, the circumference of Earth can be easily calculated. (Illustration by cmglee, David Monniaux, CC BY-SA 4.0.)

This motion tends overall in an ordered direction, but periodically reverses—called *retrograde motion*—making loops of complex shape.

Figure 2.8 shows one particular example of the path of Mars against the background of stars. The green line marks the *ecliptic*—the much simpler apparent path of the Sun against the background of stars. The ecliptic is simply a reflection of Earth's motion about the Sun, while the apparent retrograde loop of Mars arises from a combination of the motions of both Earth and Mars; we see Mars from a moving Earth.

Johannes Kepler, in the late 16th and early 17th centuries, demonstrated that all of this complexity could be explained by three elegant rules, now known as *Kepler's laws of planetary motion*. One of these (the so-called Harmonic, or Third Law) allows one to determine the size of any planet's orbit simply by measuring its period of revolution. But this size is in astronomical units—that is, it is the size *compared to* Earth's orbit. To know the actual size in kilometers, one must determine the true distance between Earth and Sun, a measurement not possible in Kepler's time.

Figure 2.8: The retrograde motion of Mars, with the planet's position marked every 30 days, is marked in red. The green line marks the ecliptic. (Sky map and planet tracks created with Cybersky, available at www.cybersky.com.)

And so from the time of Kepler it has been possible to make a scale model of the orbits of the planets. With the invention of the telescope in the early seventeenth century, it was then possible to also measure the *angular* sizes of the five naked eye planets. Galileo was the first to make hay with this; his observations of Jupiter, Saturn, and Venus as disks—not points of light—added compelling evidence to the notion that they were corporal bodies moving through space. As described in Chapter 1, Section 1.2, this gives us the ratio of the true size to the distance. And so all of the pieces were in place, by the 17th century, to make a scale model of the solar system—at least for the Sun and the five naked-eye planets, Mercury, Venus, Mars, Jupiter, and Saturn.

Apart from Uranus and Neptune, not to mention Pluto, which were discovered telescopically much later, there are two important bodies missing: Earth and Moon. The Moon does not obey Kepler's third law in the way the planets do, because it orbits Earth, not the Sun. And although the true diameter of Earth was known, its distance to the Sun was not. And so a numerical comparison between the size of Earth and the AU was unknown in Kepler's time. Kepler's contemporary, Tycho Brahe, accurately measured the distance to the Moon using parallax, but how the size of Earth fit into a scale model of the solar system had to wait until the much-later measurement of the astronomical unit.

Table 2.1: Sizes of the planets, compared to Earth. The second and third columns give, respectively, the radius and orbital size as compared to Earth. The fourth column is the angular diameter as viewed from the Sun, while the fifth column is the ratio of the body's distance to its diameter. The last two columns give scaled-down diameters and distances from the Sun, for an appropriately-scaled model. Based on numerical data from Edgar [2018, p. 22].

Body	$R/R_\oplus$	$a/a_\oplus$	θ_D (radians)	d/D	D (mm)	d (m)
Sun	109		9.3×10^{-3}	110	40	
Mercury	0.383	0.387	8.4×10^{-5}	11,900	0.14	1.67
Venus	0.949	0.723	1.1×10^{-5}	8,940	0.348	3.11
Earth	1	1	8.6×10^{-5}	11,700	0.367	4.31
Moon	0.272	1	9.3×10^{-3}	43,100	0.1	4.31
Mars	0.533	1.52	3.0×10^{-5}	33,600	0.196	6.56
Jupiter	11.2	5.2	1.8×10^{-4}	5,440	4.11	22.4
Saturn	9.45	9.56	8.4×10^{-5}	11,900	3.47	41.1
Uranus	4.01	19.2	1.8×10^{-5}	56,200	1.47	82.6
Neptune	3.88	30.1	1.1×10^{-5}	90,900	1.43	130
Pluto	0.18	39.7	3.9×10^{-7}	2,590,000	0.066	171

2.2.2   THE DATA

The second and third columns of Table 2.1 show the radii (R) and orbital sizes (a) of the Sun, Moon, and planets *as compared to the values for Earth*. The third column gives the very-tiny angular diameter, θ_D, of each planet as seen from the Sun (for the Sun and Moon it is the angular size as seen from Earth). Recall from Equation (1.11) that a small angular size means the distance is much larger than the true size of the object. This indicates an obvious basic fact about the solar system; the distances between the planets is on an enormously bigger scale than their sizes. This disparity is so great that it cannot be portrayed on a picture: if we represent the planets with dots that are big enough to see, then we need a very large sheet of paper indeed to put them at their proper distances from the Sun.

Pluto is the most extreme example. It is both the smallest planet and the farthest from the Sun. The difference between the size of Pluto and its distance from the Sun is like that of a speck of dust hundreds of feet away. Also notice that Pluto is 100 times farther from the Sun than is the closest planet, Mercury. This makes it non-trivial to draw even just the orbits of the planets on one sheet of paper to scale. Putting both the sizes of the planets and their orbits to the same scale is harder still.

The fifth column is the inverse of the angular diameter, which from Equation (1.11) is simply the planet's distance divided by its diameter. Instead of very tiny numbers, these are very large numbers; the distance of a planet from the Sun is much bigger than its diameter.

The sixth and seventh columns show an example of a particular scale model for the solar system, where I have chosen the Sun's diameter to be 40.0 mm, the size of a ping pong ball. Looking at just the data for the Sun, this sets a scale factor for our model of $s = 0.367$ scaled millimeters per Earth radius. And so our scaled diameters are given by:

$$D(\text{mm}) = 0.367 R/R_{\oplus}. \tag{2.1}$$

Our scaled distances, d, are then given simply by multiplying D (sixth column) by d/D (fifth column), except that I have converted the values from millimeters to meters, far more convenient for the much larger distances.

At a distance of 171 m from the Sun, one can certainly walk to our scale-model Pluto. The average step length is roughly 3/4 of a meter; most people can easily make 1-meter long steps if they try, but they have to stretch a bit. And so 171 meters is 171 steps with a stretch, and about 230 ordinary steps—a bit of a walk for sure, but hardly a hike. The scale-model sizes of the smallest objects in Table 2.1—the Moon and Pluto—on the other hand, are very small indeed, but should still be just barely visible to the naked eye. Jupiter, the largest planet is just a wee bit smaller than a bee-bee, and Earth, just several steps from the Sun, is about the size of a grain of salt [Beaver et al., 1996].

2.2.3   THE SHAPES OF PLANETS AND ORBITS

I have denoted the actual size of the orbits of the planets (in AU) with the symbol a, rather than "r" for radius or "d" for distance. This choice was intentional, for the orbit of a planet does not have a simple radius; planetary orbits are not circles, they are *ellipses*. An ellipse is a particular oval shape with a precise mathematical definition. There is only one thing to say about a circle—its radius. But *two* numbers are needed to specify an ellipse, as it has both a long (major) axis and a short (minor) axis. Figure 2.9 shows an ellipse, and the definition of its semi-major and semi-minor axes, indicated by a and b, respectively. Thus, the third column in Table 2.1 lists the *semi-major axis* of each planet's orbit. For completeness, I could have also listed the semi-minor axis for comparison. Or instead I could have listed the *eccentricity*—a mathematical measure of how much the ellipse deviates from a circle. But it is a that has the most relevant physical significance for the orbit, as we shall see in Part IV of *The Big Picture*.

The elliptical orbit of a planet is not centered on the Sun. Rather, the Sun is located off-center, at the *focus* of the ellipse. The origin of the coordinate axes in Figure 2.9 marks the focus of this particular ellipse, which deviates considerably more from a circle than do the orbits of any of the major planets.

I have listed the radii of the planets, with the symbol R, but that too is incomplete. To be more precise, the second column in Table 2.1 lists the *equatorial radius* of each planet. For planets

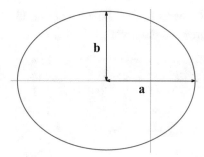

Figure 2.9: An ellipse with the semi-major axis (a) and semi-minor axis (b) marked. The origin of the coordinate axis marks the focus of the ellipse.

are not quite spheres; they are *oblate spheroids*—a sort-of squished sphere, like a basketball if you sit on it.

2.2.4 DENSITIES OF THE PLANETS

Table 2.2 shows the radius, R, volume, V, and mass, M, of the Sun, Moon, and planets, as compared to Earth. The last column shows the *density*—its mass divided by its volume.

In order to determine the mass of an astronomical object, we must see the effect its gravity has on something nearby it. For the case of the Sun, Earth and the other planets do the trick. Earth, Mars, Jupiter, Saturn, Uranus, Neptune, and Pluto all have satellites, and so the orbits of those satellites allow us to calculate their masses. The masses of Mercury and Venus, neither of which has any satellites, were not measured precisely until the Space Age, when space probes were sent past them, and the gravitational effect on these probes could be measured.

The densities in Table 2.2 are compared to water, and they are the *average* densities of these bodies. And so the average density of Earth is 5.5 times that of water. This is a density in between that of rock and metal, and that is one piece of evidence that suggests Earth is made of a combination of both—a rocky mantle and crust surrounding a metal core.

Notice that Mercury, Venus, Earth, Moon, and Mars all have densities substantially higher than that of water, while the Sun and the outer planets all have densities less than twice that of water. This is a basic fact that begs for an explanation, and we take it up in Part III of *The Big Picture*.

2.3 STARS

An ordinary star is a massive ball of mostly hydrogen and helium, held together by gravity and supporting itself by the pressure of the hot gases of which it is made. To do this, it must be hot, and the high temperature of its outer layers is the direct source of the enormous quantity of light

Table 2.2: Radii, volumes, and masses of the planets, compared to Earth, and average densities, relative to water. Based on numerical data from Edgar [2018, p. 22].

Body	$R/R_\oplus$	$V/V_\oplus$	$M/M_\oplus$	ρ/ρ_{water}
Sun	109	1,290,000	333,000	1.4
Mercury	0.383	0.0560	0.0553	5.4
Venus	0.949	0.854	0.815	5.3
Earth	1.00	1.00	1.00	5.5
Moon	0.272	0.0202	0.0123	3.3
Mars	0.533	0.151	0.107	3.9
Jupiter	11.2	1,410	317.8	1.3
Saturn	9.45	844	95.2	0.69
Uranus	4.01	64.4	14.5	1.3
Neptune	3.88	58.5	17.1	1.6
Pluto	0.18	0.0058	0.0025	1.9

that a star emits on its own. The Sun is the closest example; the rest are so distant that all but a few appear as only points of light in even the biggest of telescopes.

Table 2.3 lists size and distance data for a few stars of special significance. The sizes are listed as compared to the radius of the Sun; $R_\odot = 6.957 \times 10^8$ m. Notice the enormous range in values. Sirius B has a radius less than 1/100 that of the Sun, while Betelgeuse measures nearly 1000 solar-diameters across. This means that Betelgeuse has a radius over 100,000 times larger than that of Sirius B.

Below I briefly describe some of the basic types of stars:

- *Brown Dwarfs:* These are stars that perhaps are not even stars—it depends upon how one wants to define the term "star." The energy they emit is quite dim compared to other types of stars, and not self-sustaining. They only glow in the infrared part of the spectrum; all known brown dwarfs can only be detected with large telescopes, even though they are relatively nearby. A typical brown dwarf likely has a radius about 0.1–0.2 $R_\odot$—only somewhat bigger than Jupiter.

- *Main Sequence Stars:* The most common type of star is called a *main sequence star*. Most stars fall into this category for the simple reason that any individual star spends most of its life at this stage. Thus, we are most likely to catch a star in the act of doing just what it does for most of its life. The Sun is a main sequence star, as are the following stars in Table 2.3: Proxima Cen, α Cen A and B, and Sirius A. The Sun is a mid-sized main sequence star.

Table 2.3: Sizes of stars. The second column gives the radius as compared to that of the Sun. The third column is the distance in parsecs. The fourth and fifth columns are the star's diameter (in millimeters) and distance (in kilometers) for a scale model chosen such that the Sun is the size of a ping pong ball. Basic numerical data from Edgar [2018, p. 289].

Body	$R/R_\odot$	d(pc)	D (mm)	d (km)	d/D
Sun	1	4.85×10^{-6}	40.0	4.30×10^{-3}	108
Proxima Cen	.154	1.30	6.16	1,160	1.9×10^8
α Cen A	1.22	1.32	48.8	1,170	2.4×10^7
α Cen B	.863	1.32	34.5	1,170	3.4×10^7
Sirius A	1.71	2.64	68.4	2,350	3.4×10^7
Sirius B	.0084	2.64	0.34	2,350	6.9×10^9
Betelgeuse	900	222	36,000	197,000	5.5×10^6
Rigel	115	260	4,600	230,000	5.0×10^7

The largest have radii over 20 times that of the Sun while the smallest have radii only about 13% the Sun's.

- *Red Dwarfs:* The smallest main sequence stars are often called red dwarfs. Proxima Centauri is a good example. Although it is the nearest star to the Sun, it is much too faint to see without a telescope.

- *Blue Giants:* The very largest, brightest and hottest main sequence stars last only a relatively short time, and then they become blue giants. Rigel is a good example. It is a very hot star, much hotter than the Sun. But it is somewhat cooler than when it was a main sequence star, and it is considerably larger—over 100 times the diameter of the Sun.

- *Red Giants and Supergiants:* Red giant stars are red and, well, giant. They are red because they are relatively cool (on the outside)—still blisteringly hot, but much cooler than the Sun. Red supergiants are the largest of stars; Betelgeuse is a good example at 900 times the diameter of the Sun. From Table 2.1, Earth is about 110 solar diameters from the Sun—and so the AU is about 110 solar diameters. Since Betelgeuse is 900 times bigger than the Sun, its radius would stretch from the Sun to half way to Jupiter; Earth would be inside it.

- *Collapsed stars:* When stars get to the end of there lives, a lot can happen; we discuss these events in Part III of *The Big Picture*. But what is left over in the end is inevitably small and compact. These compact objects are very strange, in ways we will consider later, because so much mass is packed into so small of a space. There are three basic types of collapsed stars, considered in turn as follows.

- *White dwarfs:* I have already listed one example of such a compact object in Tables 2.3 and 2.4—Sirius B. It is only 0.8% the size of the Sun; in fact this is slightly smaller even than the size of *Earth*. It is very hot, and so white in color, and thus objects such as this are known as *white dwarfs*. The density is unimaginably high—over two million times that of water. No ordinary matter can achieve this density, and so it is not surprising that white dwarfs are made instead of something far more exotic—an *electron-degenerate gas*.

- *Neutron stars:* Orders of magnitude more compact even than white dwarfs are *neutron stars*. They are made of a *neutron-degenerate gas*, and a typical neutron star is only several *miles* across—about the size of a small city.

- *Black holes:* White dwarfs and neutron stars are examples of objects for which gravity is so intense that the odd effects of Einstein's general relativity (GR) become apparent. We consider these details in more detail in Part IV of *The Big Picture*, and also *black holes*—objects that are only slightly smaller than neutron stars, but far stranger.

2.3.1 A SCALE MODEL FOR THE STARS

For our smallest practical scale model of the solar system, we chose a ping-pong ball—40.0 mm in diameter—for the size of the Sun; this puts Earth at a distance of 4.31 m, and the rest of the planets at distances as given by the fifth column of Table 2.1. And so, on this same scale, how far away would be the nearest star to the Sun?

Column 5 in Table 2.3 gives the answer; three stars—all forming part of the same system in the direction of the constellation Centaurus—are about 1.3 pc away. This corresponds to a distance of nearly 1,200 km in our scale model. Notice (from the second column) that although Proxima Centauri is much smaller than the Sun, α Centauri A and B are roughly the same size. And so we have a ping pong ball Sun in, say, Appleton, WI, and a couple of other ping pong balls (one slightly larger, the other slightly smaller) 1,170 km away to represent α Centauri A and B. These two non-regulation ping-pong-ball stars would be located not *quite* so far away as Rapid City, SD, but still farther away than Wall Drug. Proxima Centauri would be just down the road from them (about 10 km closer to Appleton) but only the size of a green pea (the wrong color for this particular star). It is not at all atypical for stars that they are like ping pong balls many hundreds of kilometers apart.

2.3.2 DENSITIES OF STARS

Table 2.4 gives the radii, volumes, and average densities of the same stars as in Table 2.3. The densities, in the fourth column, are given relative to that of water, and we see that the Sun, α Centauri A and B, and Sirius A all have densities the same order of magnitude as water. But Betelgeuse and Rigel—red and blue giants, respectively—have densities orders of magnitude less than that. The opposite is true for the tiny stars. The red dwarf Proxima Centauri is 47

Table 2.4: Volumes and densities, compared to water, of examples of stars

Body	$R/R_\odot$	$V/V_\odot$	ρ/ρ_{water}
Sun	1	1	1.4
Proxima Cen	.154	0.00365	47
α Cen A	1.22	1.82	0.85
α Cen B	.863	.643	2.0
Sirius A	1.71	5.01	0.57
Sirius B	.0084	5.9×10^{-7}	2.4×10^6
Betelgeuse	900	7.3×10^8	2×10^{-8}
Rigel	115	1.5×10^6	2.2×10^{-5}

times as dense, on average, as water. And the white dwarf Sirius B has an unimaginably high density—over 2.4 million times the density of water.

In contrast to these high and low densities, the densest solid on Earth at standard temperature and pressure is osmium, with a density 22.6 times that of water. The air of Earth's atmosphere, at sea level, has a density 1.225×10^{-3} times that of water. And so from Table 2.4, the large stars Betelgeuse and Rigel both have densities orders of magnitudes *less* than air at the surface of Earth. But recall that this is the *average* density—both of these stars have far higher densities near their centers.

2.3.3 THE ANGULAR SIZES OF STARS

The last column of Table 2.3 shows the ratio of each star's distance to its diameter. From Equation (1.11), it is clear that this number is simply the reciprocal of the star's angular size (in radians) as seen from Earth. If we also apply a conversion from radians to degrees (see Section 1.2.2), we have:

$$\theta \text{ (degrees)} = \frac{57.30}{d/D}. \tag{2.2}$$

If we use the Sun's value of $d/D = 108$, for example, we have $\theta = 0.53°$. But what about for the other stars? It is clear from Equation (2.2) that a *larger* value of d/D means a *smaller* angular size. And these values are large indeed for all of the stars listed except for the Sun. Since Betelgeuse has the smallest d/D of the stars in Table 2.3, it has the *largest* angular size in the list:

$$\theta(°) = \frac{57.3}{5.5 \times 10^6} = 1.04 \times 10^{-5}. \tag{2.3}$$

And so the angular size of Betelgeuse is only one hundred thousandth of a degree! Clearly we need a smaller unit of measure, and we have already spoken of one—the arcsecond: $3600'' = 1°$.

We can then convert our angular size of Betelgeuse to arcseconds simply by multiplying degrees by 3600. This is somewhat better: $\theta('') = 0.0375$. But that is still a very small number, and so astronomers typically use the *milliarcsecond* (mas) for measurement of the angular diameter of a star other than the Sun. To convert to milliarcseconds, we simply multiply arcseconds by 1000. And so the angular diameter of Betelgeuse is about 38 mas. If we include both of these conversions into Equation (2.2), we have:

$$\theta\,(\text{mas}) = \frac{2.063 \times 10^8}{d/D}. \tag{2.4}$$

And it is not difficult to show that we can express this in terms of solar radius and distance in parsecs as follows:

$$\theta\,(\text{mas}) = 9.25 \frac{R/R_\odot}{d\,(\text{pc})}. \tag{2.5}$$

Betelgeuse is a special star because its angular size is relatively *large* compared to other stars—and yet it still appears only 38 mas across. It has, for example, about 35 times the angular size of Proxima Centauri. Proxima Centauri is the closest star to Earth and yet it has an angular diameter only a little larger than a milliarcsecond. Betelgeuse is over 200 times farther away, but its enormous size more than makes up for its greater distance. Still, Betelgeuse is about as large as a star can get, and it is also, for stars that large (which are quite rare), relatively nearby.

Consider, on the other hand, Sirius B; its distance and radius implies an angular size, from Equation (2.5), of only $\theta = 0.003$ mas. It's companion Sirius A, on the other hand, is essentially at the same distance but is 200 times bigger, and so it has a proportionally large angular diameter of $\theta = 6.5$ mas.

An angular size of 1 mas is equivalent to the apparent size of a dime, as seen from a distance of about 3700 km—roughly the distance between New York, NY and Las Vegas, NV. Is it possible to see such fine detail in even the world's most powerful telescopes? The answer is complicated, but resolving an angular size of only 1 mas is possible, but it is near the limit to what can be done with the most advanced current techniques. *And there are fewer than 20 stars with an angular size greater than 1 mas as seen from Earth.* All other stars are simply too far away, and it means that—apart from less than two dozen exceptions—stars appear as only points of light in even the biggest telescopes, and the angular size is not accurately measurable by direct observation.

2.3.4 STAR CLUSTERS

Stars are not distributed randomly in space; they tend often to be grouped together into *clusters*. Within a star cluster, stars are much closer to each other on average than they are outside of clusters. But they are still very distant from each other compared to their sizes. Star clusters play an important role in our understanding of stars. They are, in effect, nature's controlled

Figure 2.10: The open star cluster M 11, in the constellation Scutum. It is comprised of several thousand stars, measures about 40 ly years across, and lies about 6200 ly from the Sun. Image made from observations by Michael Briley, for Beaver et al. [2013].

experiments—locations in space where many stars formed out of the same cloud of gas and dust, and all roughly *at the same time*. We will consider them often throughout *The Big Picture*.

Open Clusters

Open star clusters are also called *galactic star clusters* because they tend to be located along the Milky Way. They consist of dozens to many thousands of stars. A good example is M 11, located in the constellation Scutum; it appears in a very small telescope as a dim cloud with an angular diameter of about 25 arcmin. In a larger backyard telescope it can be seen as a scattering of hundreds of stars, while about three thousand can be detected with large telescopes. See Figure 2.10 for an image made with a telescope at Kitt Peak National Observatory, as part of the data set for Beaver et al. [2013].

This cluster is estimated to be about 6000 ly distant [Beaver et al., 2013]. We can use this fact, in combination with its observed *angular* size, in order to determine its true size in space, as described in Section 1.2. If we multiply the *angular* size of M 11 by its distance, we can calculate

Figure 2.11: The globular star cluster M 13 in the constellation Hercules. It is comprised of several *hundred* thousand stars, measures about 130 ly across, and lies about 22,000 ly from the Sun. (Image by Rawastrodata, CC BY-SA 3.0.)

its actual diameter in light years. We must first convert the 25 min of arc to radians:

$$D \;=\; \theta_D d \tag{2.6}$$
$$D \;=\; 25' \left(\frac{1°}{60'}\right) \left(\frac{0.01745\,\text{radians}}{1°}\right) 6000\,\text{ly}. \tag{2.7}$$
$$D \;=\; 44\,\text{ly} \tag{2.8}$$

And so this particular cluster extends across a volume of space some 44 ly in diameter. There are dozens of these open star clusters visible in our region of the Galaxy. M 11 is a particular rich example, and contains considerably more stars than the average open cluster.

Globular Clusters

Figure 2.11 shows one of the most famous examples of a *globular cluster*, as imaged by a small backyard telescope. These clusters contain up to many hundreds of thousands of stars—in some cases over one million. They are roughly spherical in shape and are almost always much farther away than the open clusters we see.

M 13 is over 22,000 ly away, and yet it appears almost as big in the sky as M 11, even though it is more than three times farther away. This means M 13 must be a few times larger

Figure 2.12: **Left:** The Orion Nebula, in the "sword" of the constellation Orion, is a glowing cloud of ionized hydrogen. Photo by the author. **Right:** The constellation of Orion lies in the direction of a giant molecular cloud, of which the Orion Nebula is only a tiny part. (Image made with the Aladin sky atlas. DSS2, alasky.u-strasbg.fr/DSS/DSSColor [Bonnarel et al., 2000, Lasker et al., 1996].)

than M 11. But when we consider that this globular cluster contains 300,000 stars—one hundred times as many as M 11—it is clear that inside a globular cluster the stars are much closer together.

2.4   HII REGIONS AND GIANT MOLECULAR CLOUDS

The left side of Figure 2.12 shows a small part of the constellation Orion—the faint "sword" just below the three stars of the prominent "belt"—as imaged with a long exposure through a small telescope. The glowing red gas is the Orion Nebula, a region of ionized hydrogen, also called an *HII region.*

The right-hand image in Figure 2.12 is a mosaic of images made with a large telescope, and it shows that nearly the entire constellation of Orion is swathed in a complex of both dark and glowing gas and dust. This *giant molecular cloud* (GMC) is about 240 ly across. At a distance of less than 1500 ly, it covers over 10° of the sky. The Orion nebula can be seen as the small bright patch just below the three stars that make up the belt of Orion. A typical GMC is about 150 ly across [Carroll and Ostlie, 2017, p 407], and so at about 240 ly the Orion GMC is a big one.

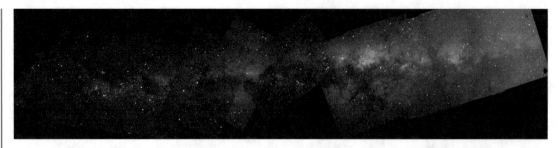

Figure 2.13: A mosaic of 12 photographs showing the Milky Way (as seen from the Southern Hemisphere) stretching across the sky from horizon to horizon. Photograph by the author.

The HII region we see as the Orion Nebula is much smaller than the GMC it is part of. The light comes from ionized hydrogen gas, and the energy for the ionization comes from ultraviolet light produced by hot type O and B stars. And so the size of the glowing region is limited by the ultraviolet light emitted by such stars; an HII region may be less than a light year across. But since multiple stars can contribute to the ionization, the largest HII regions are a few hundred light years across [Carroll and Ostlie, 2017]. The Orion Nebula is about 20 ly across.

Most of the Orion GMC appears dark at visible wavelengths. The gas is relatively dense, and very cool—typically only about 15 K. The low temperatures mean that molecules can easily form. The visible HII regions scattered throughout the Orion GMC are places where newly formed O and B stars are ionizing the gas, and thus raising it to temperatures above 10,000 K. The red color comes primarily from emission of hydrogen.

The Orion Nebula will eventually become an open star cluster. Indeed, there already is a newly-formed open star cluster inside it. But the presence of dust obscures our view at visible wavelengths. The gas and dust in between the stars, including HII regions and GMC's but also less-dense material, is called the *interstellar medium* (ISM).

2.5   GALAXIES

The word galaxy comes from the Greek word *galaxias*, which means "milky." This refers to the dim, milky glow of light that forms a band stretching across the nighttime sky—the Milky Way. We now know this is our own galaxy as seen from the inside, and it is only one of billions of galaxies in the universe. See Figure 2.13 for my own photograph of the Southern Hemisphere Milky Way, made from a mosaic of 12 photographs, stretching clear across the sky from horizon to horizon.

2.5.1   THE MILKY WAY AND THE ANDROMEDA GALAXY

"The Milky Way" is the name for both the galaxy we are part of, and also its visual appearance in the nighttime sky. It is shaped like a flat disk with a central bulge, made of several hundred

Figure 2.14: A panorama of the Milky Way. We are seeing it from the inside, and we have a view that sees the disk of the galaxy edge-on. (Image by ESO/S. Brunier, CC BY 4.0.)

billion stars. The Milky Way stretches more than 100,000 light years across, and is about 1000 light years thick (thicker at the bulge). That is, the Galaxy is so big that it takes light 100,000 years to travel from one side to the other.

The disk is where most of the stars and nearly all the gas and dust reside, but the Milky Way also has a *halo*—a spherical distribution of stars that surrounds the disk and extends to distances beyond its edge. The globular clusters are part of the halo of the Milky Way, while the open clusters are part of the disk.

Figure 2.14 shows a photo-montage of the Milky Way, made by the European Southern Observatory from many individual images stretching nearly around the entire sky. The glow is made from the combined light of millions of stars too faint to show individually. The dark patches are dust, concentrated in the disk of the Galaxy, that obscures the view of stars beyond.

We see other galaxies besides our own, very far away, and some look just like we would expect the Milky Way to appear if seen from afar. The galaxy NGC 891 (see Figure 2.15) in the constellation Andromeda is a good example.

The constellation Andromeda holds one of the most famous of galaxies—the Andromeda Galaxy, M 31. It is much closer than NGC 891, and appears somewhat as we expect the Milky Way would appear if seen with the disk slightly inclined. M 31 is easily visible from the Northern hemisphere on a dark night with the naked eye. It is probably about 40% larger than our Milky Way, and it is the nearest large galaxy. With a diameter of 70 kpc and a distance of 0.78 Mpc, we can easily calculate its size.

Figure 2.15: We see other galaxies that look like our Milky Way. This is NGC 891, and it is a spiral galaxy like our own, seen edge-on. (Image by Hewholooks, CC BY-SA 3.0.)

Here I have referred to the common units of measure used by astronomers to describe the sizes and distances of galaxies. We have already described the parsec (pc) to measure the distances between stars within the Milky Way. But the size of a typical large galaxy like our own suggests a more convenient unit, the kiloparsec (kpc). One kiloparsec is equivalent to 1000 parsecs.

For the vast distances between galaxies, the *megaparsec* (Mpc) is more convenient. A megaparsec is equivalent to a million parsecs or 1000 kpc. And so let us look again at the fact that the Andromeda galaxy is 70 kpc across and 0.78 Mpc distant. If we rewrite 0.78 Mpc as 780 kpc, it is clear that the angular diameter of the Andromeda galaxy—its diameter divided by its distance—is 70/780 = 0.009 radians. This is equivalent to about 5°, which is about 10 times larger than the apparent size of the full Moon. But when observed directly, the Andromeda galaxy appears as a dim cloud of low surface brightness; good eyes and extremely dark and clear skies are required to visually trace its outer environs. Long exposure photography, on the other hand, can easily reveal its beauty; see Figure 2.16.

Figure 2.16: The Andromeda Galaxy, M 31, is the closest large spiral galaxy to our own Milky Way. (Image by Adam Evans, CC BY 2.0.)

2.5.2   DWARF GALAXIES

Notice that there are two small blobs of light next to the Andromeda Galaxy, in Figure 2.16. These are the *dwarf elliptical galaxies* M 32 and NGC 205. Dwarf galaxies are sometimes irregular in shape (dwarf irregulars) or more elliptical or spheroidal (dwarf ellipticals), and they make up the vast majority of galaxies by number. A good example of a dwarf irregular is the Small Magellanic Cloud (SMC), easily visible to the naked eye from the Southern hemisphere. It is about 61 kpc distant and roughly 2 kpc across. Compare this to the Milky Way, which is over 30 kpc across.

Next to the SMC in the sky is the Large Magellanic Cloud (see Figure 2.17). It is a bit closer than the SMC, and about twice its size. There is evidence that it is not really irregular; rather it is the remains of a barred spiral galaxy, distorted by the gravity of the SMC. And so it is often considered a peculiar and very small barred spiral galaxy, rather than a dwarf irregular galaxy. Both the SMC and LMC contain on the order of 10 billion stars—far fewer than the roughly 250 billion that make up the Milky Way.

Figure 2.17: The Large Magellanic Cloud (LMC) and the Small Magellanic Cloud (SMC) as photographed from near Casablanca, Chile. They can be seen as the two fuzzy blobs near the center top of the image. The Milky Way stretches downward from the upper left. Photograph by the author.

2.5.3   GIANT ELLIPTICALS

The largest galaxies are the giant elliptical galaxies. Both giant and elliptical at the same time, these behemoths may contain over 1 trillion stars. As is the case with most ellipticals, they are mostly stars, with very little gas or dust. A good example is the galaxy M 87 in the Virgo Cluster of Galaxies. At a distance of 16.4 Mpc, it is nearly 300 kpc across, six times the diameter of the Milky Way. M 87 is at the center of a large cluster of galaxies called the Virgo Cluster (see Section 2.6). It is not uncommon for a giant elliptical to lurk at the center of a large cluster of galaxies.

2.6   CLUSTERS OF GALAXIES

Like stars, galaxies are often distributed in space in clusters, rather than simply randomly scattered. Galaxy clusters are often given names corresponding to the constellation they appear to

Figure 2.18: The giant elliptical galaxy M 87. (Image credit: Nasa/STScI/Wikisky, Public Domain.)

lie within. Of course the constellations are simply chance groupings of stars comparatively quite near to us; the galaxies are many orders of magnitude more distant.

2.6.1   THE LOCAL GROUP

Our own Milky Way galaxy is in a small cluster called the *Local Group*. It consists of two large spiral galaxies—the Milky Way and the Andromeda Galaxy—along with another medium-sized spiral galaxy (M 33) and many dozens of dwarf irregulars and dwarf ellipticals. The Local Group is roughly 3 Mpc across.

2.6.2   THE VIRGO CLUSTER

The nearest large cluster of galaxies is the Virgo Cluster. It contains some 1500 galaxies, probably many more if we could detect all of the faint dwarf galaxies presumably contained within. The largest is the giant elliptical M 87, which lies near the cluster's center. The center of the Virgo cluster is about 16 Mpc distant, and it extends to a diameter of roughly 4.5 Mpc.

2.6.3   SUPERCLUSTERS AND VOIDS

The Local Group, the Virgo Cluster, and a few other small clusters of galaxies are known to be physically bound together because of the mutual gravity they feel for each other. Together they make up what is known as the Local Supercluster, or the Virgo Supercluster. And so a supercluster can be thought of as a cluster of clusters of galaxies.

Between the superclusters are large regions that are relatively free of galaxies. These are called *voids*, and they are roughly spherical in shape. Superclusters on the other hand are often shaped like elongated filaments or flattened sheets. The typical size of both superclusters and voids is roughly 100 Mpc; they are the largest structures in the universe [Ryden, 2017, pp. 10, 209].

2.7   CONSTELLATIONS AND THE VIEW FROM EARTH

The traditional patterns of stars visible to the naked eye are called *constellations*. We have no depth perception when looking at the celestial sphere, so the constellations are essentially two-dimensional patterns that tell us nothing *per se* about the three-dimensional arrangement of the stars.

It turns out that the visual appearance of most constellations is wildly misleading. Many of the brightest stars visible to the naked eye are quite distant, while most of the nearest stars are too faint to see without a telescope. Some stars emit more light than others, by many powers of ten—and so just because a star appears bright, it does not mean that it is relatively nearby.

And so most of the constellation patterns are chance arrangements of stars that simply happen to be roughly in the same direction, but at vastly different distances. There are a few exceptions however. The seven bright stars that form the familiar pattern of the Big Dipper in the Northern sky, for example, really are relatively close to each other in space. They form what is essentially a very sparse, nearby open star cluster. The same is true for many of the bright stars in Orion, and the V-shaped head of Taurus the Bull.

If we include other astronomical objects besides stars, the appearance in the night-sky can be even more misleading. Figure 2.19 is a time exposure, with an ordinary camera lens, of comet Hale–Bopp, which graced the skies in 1997. The trees in the foreground are blurred because the camera was tracked to counteract Earth's rotation, and so track with the stars during the roughly 15-min exposure.

The comet was roughly 1 AU from Earth when this picture was taken; the stars, of course, are in the background, many light years distant. The visible comet was nothing but sunlight reflected off a large but insubstantial cloud of dust (the white part) and glowing gas (the blue part) that stretched tens of millions of miles. But the solid part of the comet—its nucleus, hidden in the depths of the dust cloud—was only some 60 km across.

Figure 2.19 also shows another object. The faint, somewhat fuzzy blob to the lower left of the head of the comet is the Andromeda Galaxy, some 2.5 *million* light years distant.

Figure 2.19: Comet Hale–Bopp, the Andromeda Galaxy (the small diffuse patch to the lower left of the comet), and parts of the constellations Cassiopeia and Andromeda. Photograph by the author.

2.8   FROM THE MILKY WAY TO 3C 273

Quasars are some of the most distant objects observed. The nearest is 3C 273, and it appears as a dim star in a backyard telescope. With spectroscopic analysis, it is clear that this is an object emitting enormous quantities of energy, and it is at the vast distance of 750 Mpc. In Table 2.5, I list the diameters and distances of a range of galaxies extending form our Milky Way to 3C 273. The galaxies are chosen as representative members of several galaxy clusters.

Note that the largest galaxy listed is M 87, at 300 kpc. The largest distance is 750 Mpc, or 750,000 kpc. Thus, 3C 273 is located about 2,500 M 87-diameters distant. These numbers suggest we can easily construct a useful scale model to illustrate this range of distances and sizes for galaxies. Such a scale model was impractical for the individual stars within our own Milky Way; they are like ping pong balls spaced hundreds of kilometers apart. Although galaxies are

Table 2.5: Some representative galaxies between the Milky Way and the distant quasar 3C 273. The last column gives the name of the galaxy cluster of which each is a member.

Galaxy	Diameter (kpc)	Distance (Mpc)	Cluster
Milky Way	50	0	Local Group
Andromeda Galaxy	70	0.78	Local Group
M 33	20	0.81	Local Group
NGC 300	46	2.15	Local Group
NGC 55	20	2.17	Local Group
M 81	28	3.70	M81 Group
M82	11	3.59	M81 Group
M87	300	16.4	Virgo
NGC 1399	40	19	Fornax
ESO 137-001	30	68	Norma
NGC 4911	40	98	Coma
NGC 6041	67	145	Hercules
Hoag's Object	40	188	
LEDA 20221	300	330	Gemini
LEDA 51975	300	570	Bootes
3C 273	60	750	

vastly more distant from each other than are the individual stars within our Milky Way, the enormous size of a typical galaxy more than makes up for this.

Let us draw our scale model on a typical 75-m long roll of 40-mm wide, cash register receipt tape. If we pick our scale such that 300 kpc is equivalent to 30 mm on our tape, then a drawing of our largest galaxy will fit with only a little room left over. The scale is then 1 mm on our tape to represent 10 kpc in space. Since there are 1000 mm in a meter, and 1000 kpc in a Mpc, it also means that one meter of our tape represents 10 Mpc of distance in space. And so for this choice of scale we have the simple result that the numbers, divided by ten, in the second column of Table 2.5 represent the diameter of each galaxy in millimeters on our tape. And the numbers in the third column—also divided by ten—represent the scaled distance of each galaxy from the Milky Way, in meters.

The smallest galaxy on this scale is only a little over 1 mm across—tiny but still easy to see (or draw). The total distance can be easily walked. See Figure 2.20 for an example.

Figure 2.20: A scale model on cash register tape. The Milky Way is at one end, and the galaxy LEDA 20221, 330 Mpc away, is near the other end of the 40-m long tape. At this scale, a giant elliptical galaxy such as M 87 just fits across the tape's 40-mm width. (Photograph by the author.)

2.9   THE DEEP FIELD

The Hubble Ultra Deep Field (UDF) is a photo montage taken with the Hubble Space Telescope; see Figure 2.21. Nearly every one of the many thousands of objects on the image is a distant galaxy. Only a few of the objects are nearby stars within our own Milky Way; they are easily identified by the cross-like diffraction spikes created by the secondary mirror supports within the telescope.

The UDF covers a very tiny piece of the sky—only a few arcminutes wide—essentially peering between the stars in our own Milky Way. The most distant objects visible in the image are estimated to be nearly 4,000 Mpc (13 billion light years) away, and so the UDF show us a part of the universe less than one billion years after the Big Bang (see Part II of *The Big Picture*).

2.10   THE END OF SPACE

We see back in time when we look out into space. And so if there was a *time* before which there was nothing to see, it follows that there is a maximum distance that we can see in space. In Section 5.4.2 we will consider the Big Bang—the notion that the universe was once extremely hot and dense, and has been expanding and cooling since. Early in the Big Bang, the entire universe—although full of photons of light—was opaque to their uninterrupted passage through

Figure 2.21: The Hubble Ultra Deep Field. The enlarged detail on the right is of the area boxed in red in the full image on the left. Only a few Milky Way stars show in the full image, identifiable by the cross-like diffraction pattern (from the secondary mirror support within the telescope) they display. The other ~10,000 objects in the image are galaxies. (Image credit: NASA, ESA, and S. Beckwith (STScI) and the HUDF Team, Public Domain.)

space. And so we can look back no further than to the time, over 13 billion years ago, when the universe first became transparent.

The light from that *epoch of recombination* is called the *cosmic microwave background*, and it tells us what the universe was doing about 13.5 billion years ago, when all of everything was still too hot to touch. This cosmic microwave background (CMB) can be seen in every direction with microwave telescopes. Figure 2.22 shows a map of the CMB covering the entire sky, from the Wilkinson Microwave Anisotropy Project.

The CMB is important because it shows us the beginnings of large-scale structure in the universe. And this structure—concentrations of mass held together by gravity—is responsible for all that we know, ourselves included. We consider this topic further in the last chapter of *The Big Picture*.

2.10.1   THE COSMOLOGICAL HORIZON

And so how far away *is* the CMB? There is no simple answer; indeed the question itself is more complex than one might think. If the Big Bang occurred 13.8 billion years ago, then one might expect that we can see objects no more distant than 13.8 billion light years away—the distance

Figure 2.22: A map of the cosmic microwave background as made by the WMAP project. (Image by NASA/WMAP Science Team - NASA/WMAP Science Team, Public Domain.)

light travels in 13.8 billion years. But it is not so simple as this because our universe is dynamic, not static—it is *expanding*. The finite age of the universe does still mean, however, that there is a maximum distance we can see. For objects more distant than this *cosmological horizon* or *particle horizon*, light would not yet have had enough *time* to reach us. From current models, the cosmological horizon is estimated to be roughly 14,000 Mpc distant—about 46 billion light years away [Ryden, 2017, p. 98].

2.11   REFERENCES

John Beaver. *The Physics and Art of Photography, Volume 1: Geometry and the Nature of Light.* IOP Publishing, 2018. DOI: 10.1088/2053-2571/aae1b6 28

John Beaver, W. Scott Kardel, and Greg Novacek. Scaling the solar system at lake afton public observatory. In John R. Percy, Ed., *Astronomy Education: Current Developments, Future Coordination*, p. 167, Astronomical Society of the Pacific, 1996. 34

John Beaver, Nadia Kaltcheva, Michael Briley, and Dan Piehl. Strömgren H-β photometry of the rich open cluster NGC 6705 (M 11). *Publications of the Astronomical Society of the Pacific*, 125(934), 2013. DOI: 10.1086/674175 41

F. Bonnarel, P. Fernique, O. Bienaymé, D. Egret, F. Genova, M. Louys, F. Ochsenbein, M. Wenger, and J. G. Bartlett. The ALADIN interactive sky atlas. A reference tool for identification of astronomical sources. *Astronomy and Astrophysics Supplement*, 143:33–40, April 2000. DOI: 10.1051/aas:2000331 xxii, xxiii, 43

Bradley W. Carroll and Dale A. Ostlie. *An Introduction to Modern Astrophysics*, 2nd ed., Cambridge University Press, 2017. DOI: 10.1017/9781108380980 43, 44

James S. Edgar, Ed. *Observer's Handbook 2019*. The Royal Astronomical Society of Canada, Toronto, 2018. 33, 36, 37

B. M. Lasker, J. Doggett, B. McLean, C. Sturch, S. Djorgovski, R. R. de Carvalho, and I. N. Reid. The Palomar—ST ScI Digitized Sky Survey (POSS–II): Preliminary Data Availability. In G. H. Jacoby and J. Barnes, Eds., *Astronomical Data Analysis Software and Systems V*, volume 101 of *Astronomical Society of the Pacific Conference Series*, pp. 88, 1996. xxii, xxiii, 43

Barbara Ryden. *Introduction to Cosmology*. Cambridge University Press, 2017. 50, 55

CHAPTER 3

Looking Inward

3.1 SELF GRAVITATION

The Moon, made mostly of rock, is spherical in shape. And yet, pick up a random rock here on Earth and it is very unlikely to be even approximately a sphere. A typical rock is, well, hard as a rock; but the Moon is harder still. The difference arises because the rock you hold in your hand is held together by chemical bonds, while the Moon is held together by gravity. Every part of the Moon is attracted gravitationally to every other part, and it is this *self gravity* that pulls the Moon into a spherical shape. The reason the rock does not do the same is simply that it has so little mass, and thus only a minuscule amount of self-gravity.

With this realization comes an obvious question: how much mass must an object have in order for its own gravity to hold it together and pull it into a sphere? And how large would the smallest such object be? We can get some hint at this answer by looking at different solid objects in the solar system. All of the planets are self-gravitating spheres, but the asteroids and some of the smaller satellites of the planets are not. They are instead a variety of random rock-like shapes.

The smallest self-gravitating bodies in our solar system are roughly 500–1000 km in diameter. Bodies larger than that have enough mass, and thus self gravity, to pull themselves into a spherical shape. Bodies that are much smaller on the other hand, are random rock shape. They are held together just like ordinary rocks—by the intermolecular forces between its components. These smaller bodies include the comets, most of the asteroids (rocky bodies mostly between the orbits of Mars and Jupiter) and some of the smaller moons of the planets.

3.2 THE SIZE OF LIFE

Depending upon how one defines "life," the smallest living organisms are bacteria (about 10^{-6} m), viruses (about 10^{-7} m), or prions (about 10^{-8} m). The blue whale, on the other hand—thought to be the largest organism ever to have lived—is about 30-m long. And so the 12,800 km diameter Earth is about 430,000 blue whales across. Sadly, this is an order of magnitude more blue whales than exist on our planet; they were hunted nearly to extinction in the first half of the twentieth century. We can also see that the length of a blue whale is about eight orders of magnitude greater than the diameter of a virus. This is similar to the distance of the nearest star, Proxima Centauri, when compared to its size.

3.3   THE MICROSCOPIC

An ordinary microscope is an arrangement of lenses that allows one to directly see a magnified view of tiny objects. No matter how perfect the optics, there is a limit that cannot be surpassed; one cannot use an ordinary microscope to see anything smaller than the wavelength of light itself. For humans, this wavelength is roughly 5×10^{-7} m, just a bit smaller than the smallest bacteria. And so objects smaller than this must be "viewed" by other means.

3.4   MOLECULES, ATOMS, AND THEIR PARTS

Individual atoms are roughly 10^{-10} m in size, about 4 orders of magnitude—10,000 times—smaller than the smallest objects visible through microscopes. Atoms are made of a nucleus of protons and neutrons, surrounded by a cloud of electrons. The electrons (with a negative electric charge) have very little of the overall mass, *but they take up nearly all of the space of an atom.* And so it is the electrons that give the atom its size.

The nucleus of an atom, on the other hand, made of protons and neutrons, makes up most of the atom's mass but takes up very little of the space. A typical atomic nucleus is only 10^{-19} m across, nine powers of ten smaller than the atom itself. This is similar to the size of Earth compared to about one fifth of the *distance* to the nearest star, α Centauri. And so, borrowing from our scale model of the solar system, the nucleus of an atom is like a grain of salt in an atom that is nearly 250 km across.

Molecules are arrangements of atoms bound to each other by electrical forces. A water molecule, made from two atoms of hydrogen and one of oxygen, is a familiar example. It is not much larger than a single atom of oxygen, but molecules exist that are made of many thousands of atoms. The large rod-shaped protein molecule *fibrogen*, for example, is 50 nm (5×10^{-8} m) long [Erickson, 2009]. This is over two orders of magnitude larger than a single atom, but still far too small to see with a microscope.

3.5   THE PLANCK LENGTH

There is a smallest size where *we know that we don't know* what is going on; the laws of physics as presently understood simply do not hold at this scale. It can be calculated from simple assumptions, combining the known physical constants in such a way that they make a length. Thus, our constants that describe the speed of light (c), the strength of gravitation (G), and the fundamental quantum of physics, Planck's constant (h) form a length, l_P, when combined as follows [Ryden, 2017, Chap. 1]:

$$l_P = \sqrt{\frac{Gh}{c^3}}. \tag{3.1}$$

This length is very tiny, and it is known as the Planck length, after the physicist Max Planck. The Planck length is about 1.6×10^{-35} m, and at lengths smaller than that, our current understanding of physics breaks down [Penrose, 2004, p. 872].

3.6 REFERENCES

Harold P. Erickson. Size and shape of protein molecules at the nanometer level determined by sedimentation, gel filtration, and electron microscopy. *Biological Procedures Online*, 11(1):32–51, May 2009. DOI: 10.1007/s12575-009-9008-x 58

Roger Penrose. *The Road to Reality: A Complete Guide to the Laws of the Universe*. Vintage Books, 2004. 58

Barbara Ryden. *Introduction to Cosmology*. Cambridge University Press, 2017. 58

PART II

Time

CHAPTER 4

Tools for Understanding Time

4.1 TIMELINES

A history is a sequence of events that happen over time. The difficulty is that important things happen over both very short and very long time intervals. One way to deal with this is to make a *timeline*—we convert time into a convenient *length* scale that is easy to experience directly.

A 70-m long roll of cash register receipt paper makes for an excellent medium with which to embody a timeline. As an example, let us imagine using such a roll of paper to make a timeline of an entire human life of 100 years. If we use the full 70 m length of the roll to represent 100 years, then what is the *shortest* amount of time that we can record in practice? Assuming that we could make marks on the paper as small as 1 mm apart, our full roll of paper is thus 70,000 such millimeters long. And so we could make marks to represent a time span as short as $(100/70,000)$ yr $= 0.00143$ yr ≈ 12.5 hr.

See Figure 4.1 for a tiny detail of the timeline of my life. And so at a scale of 1 mm for every 12.5 hr, our timeline reaches an easily-walkable 70 m to represent the time span of 100 years.

But as is the case for a scale model, timelines are of limited use for conceptualizing time scales that include important intervals that vary over many powers of ten. And so our 70-m roll of paper can easily be used to encompass a century. But at such a scale, it is useless for portraying events that happen on scales of minutes, seconds or fractions of a second.

4.1.1 LOGARITHMIC TIMELINES

We can easily walk the length of a paper tape that is, for example, 100 m long. And if laid out in a large field, we could look over the entire length of the tape, and so directly "experience" its length. The *smallest* detail we can directly experience, however, is on the order of 1 mm. This would give us a factor of about 100,000—five powers of ten—between the largest and smallest times we could record on our timeline.

As we shall see in Chapter 6, a history of the universe includes important events that occur over time intervals that vary by *sixty powers of ten*. And so it would be pointless to make proportional marks on a paper tape as a visual representation of key events in the history of the universe. Nearly all of the marks would appear all bunched up on top of each other at the beginning of the tape. We can, however, modify our visual timeline through use of the mathematical tool of *logarithms*, discussed in Chapter 1.

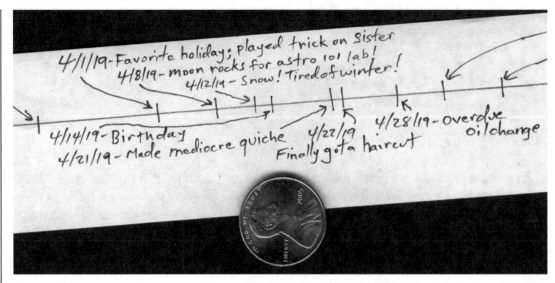

Figure 4.1: A small percentage of the timeline of my life. At the chosen scale, if I live to be 100 years old it will all fit on a single 70-m roll of register receipt paper.

Figure 4.2: A timeline of the events in Table 4.1 can easily be incorporated in a small space if a logarithmic scale is used instead of a linear one. But it is more difficult to intuitively grasp the result.

As an example, consider the list of times in the first column of Table 4.1. These particular times mark key events in the history of the universe, as we shall describe in more detail in Chapter 6. The second column gives scaled distances for our timeline, such that the longest time just fits on our 100-m long paper tape. Notice that the first four marks are all microscopically close to zero, on our 100-m long tape. Indeed, that is an understatement; 4.2×10^{-14} is three orders of magnitude smaller than a hydrogen atom, and 2.3×10^{-16} m is smaller than the dimensions of an atomic nucleus! Even our fifth mark is only a mere 2.8 mm from our zero mark.

The third column is the base ten common logarithm of the time t, and the fourth column represents an appropriately scaled timeline for these logarithms. The logarithmic scale does so well at compressing many powers of ten that we clearly need not bother with a 100-m-long paper tape. In fact the entire logarithmic time line will fit on the page of this book, although the spacing between the last two time events is barely visible at this small scale. See Figure 4.2.

Table 4.1: The times of some key events in the history of the universe and a scaled 100-m timeline, on both a linear scale (second and third columns) and a logarithmic scale (fourth and fifth columns)

Event	$t(s)$	$x(m)$	$\log t$	$x'(m)$
End of Planck era	1.0×10^{-43}	2.4×10^{-59}	-43	0
End of inflation era	1.0×10^{-32}	2.3×10^{-51}	-32.0	18.2
Proton-neutron ratio fixed	1.00	2.3×10^{-16}	0.00	71.0
End of fusion era	180	4.2×10^{-14}	2.26	74.7
Time of decoupling	1.2×10^{13}	2.8×10^{-3}	13.1	92.5
Re-ionization begins	2.1×10^{16}	4.88	16.3	97.9
Formation of Milky Way	2.4×10^{17}	55.8	17.4	99.6
The present	4.3×10^{17}	100	17.6	100

Although a logarithmic time line easily accommodates both tiny and enormous times, it is less intuitive. One needs to first learn to "think logarithmically" in order to understand Figure 4.2 at a glance.

4.2   LIGHT-TRAVEL DISTANCE

We have already seen that one can portray enormous distances in terms of the time light is required to travel. We can do this same trick in reverse in order to more easily conceptualize unimaginably small intervals of *time*. We can express time in terms of the *distance* light travels during that time. For example, light travels 30.0 cm in 10^{-9} s (one nanosecond). It is not possible to directly experience or even imagine a billionth of a second. But the light-travel distance—30 cm—can literally be held in one's hand. See Figure 4.3.

Our concept of a light-travel distance to express time is similar to our Chapter 1 description of light-travel time to indicate distance. Light travel time was a useful concept especially for very large distances. The opposite is true here; light travel distance is useful mostly for describing very tiny intervals of time.

4.3   LOOK-BACK TIME

Because of the particular finite speed traveled by light—30,000 km/s or 30 cm/ns—we never see the present. Instead, we look into the past. As I write this, Tobias the cat sits about 30 cm in front of me, trying to block my view of the computer monitor. But I see him not as he is, but rather how he was about one billionth of a second ago.

Figure 4.3: My personal nanosecond, cut to 30.0 cm from a polycarbonate rod and proudly displayed on its oak stand. Tail of cat (Tobias) is shown for scale.

This *look-back time*—extremely tiny for the distances of our ordinary experience—has profound implications for astronomy. It means that when we look out into space, we look back in time. If we express astronomical distances in light years, then the look-back time is simply that number of years. The star α Centauri A is 4.3 light years distant, and so its look-back time is simply 4.3 years.

This begs an obvious question: since we see only the past, when we look up at the stars in the night sky are they—right now—markedly different from how we see them? For our naked-eye view, the answer is likely to be, "not all that much." Yes, we see the stars not as they are, but rather how they were years, dozens of years, centuries and even thousands of years in the past. But as we will see in Chapter 9, even a thousand years is a blink of the eye for the vast majority of stars.

Look-back time allows us to survey the history of the universe, simply by looking farther and farther into space. And so it is a crucial tool of cosmology. But because of the finite age and expansion of the universe, and the effect matter has on the geometry of space and time, the relation between look-back time and distance is more complex for extremely distant objects [Ryden, 2017, p. 98].

4.4   THE CYCLE OF TIME AND THE ARROW OF TIME

Some physical process repeat themselves on a regular schedule, like the vibration of a spring, the rotation of Earth on its axis, or Earth's orbit about the Sun. To the extent that these cyclical process are regular, they can be used as a way to measure time simply by counting. And this is in fact how the second is defined by physicists—a particular agreed-upon number of oscillations of a particular atom under particular conditions. Nearly all clocks work on the same principle as well; some physical processes causes an oscillation, and we simply count repetitions of that natural cycle.

Light itself is a good example; it is a wave of electricity and magnetism that passes through space. As a light wave passes by a particular point, the electric and magnetic properties of space change rapidly back and forth, many times per second. This *frequency* is measured in repetitions per second, or hertz (Hz). And so $1\,\text{Hz} = 1\,\text{s}^{-1}$. For the light that is visible to the human eye, this frequency is on the order of 10^{15} Hz, an unimaginably high frequency—and so every cycle passes in an unimaginably *tiny* amount of time.

But this is not the only way to look at time, and a sand or water clock is an obvious counterpoint. The sand runs steadily through the tiny opening, at a regular rate, and time is measured by the quantity of sand that has run out. Instead of a repeating cycle, there is a steady progression in a particular direction.

This so-called *arrow of time* arises because many physically complex systems naturally progress from an ordered to a more disordered state. Consider a glass that falls off the table onto the floor; the ordered structure is pulverized into random pieces. Now imagine the process reversed in time. We would instantly recognize it as fanciful—pieces of glass do not magically jump off the floor onto the table and spontaneously assemble themselves into a glass.

At a more microscopic level, imagine a bit of gas released into one corner of an evacuated container. The *random* process of the molecules colliding with each other would quickly cause the gas to diffuse evenly throughout the enclosure. There are many more directions away from the initial release point than toward it.

This is the foundation behind the important physical concept of *entropy*, and there is a sense in which this progression is inevitable simply due to the laws of probability. But many physicists think there may be more to it than that, and the deep connection between the arrow of time and the cycle of time is still a matter of debate; for example, see Sachs [1987] and Penrose [2004, Chap. 27 and Sec. 30.3].

4.5   THE DOPPLER EFFECT

Motion connects time to space, and the electromagnetic wave motion of light is of particular importance. We will talk more about waves, and light in particular, in Chapter 12, but here we note an important consequence of motion for even the most basic observations of light.

We can describe light in terms of the frequency of its wave, the inverse of the time required for a particular repetition of the wave to pass a point in space. But we can also consider the *distance in space* over which the wave repeats itself at a particular instant of time, and this is called, appropriately enough, the *wavelength*, usually denoted by the Greek letter lambda (λ). Since it is a length, it has ordinary units of meters.

Because speed is distance per time, the frequency, f, wavelength, λ, and speed, v, of any wave are necessarily connected to each other:

$$v = f\lambda. \tag{4.1}$$

But light travels (in a vacuum) always at the same speed, c, and so for light, $c = f\lambda$. Imagine that an electromagnetic wave, emitted by some source, approaches you—at the speed of light—while you are moving *towards* the source. You will pass through each wavelength at a greater rate than if you and the source were stationary relative to each other. And thus you would observe a *higher* frequency. Similarly, if you were moving away from the source of light, you would measure a *lower* frequency for the incoming light.

This result is known as the *Doppler effect*, and for most waves the mathematical connection depends on the motions of the wave, the source and the observer relative to the wave medium. But for light, it is both simpler and more complex.

It is simpler because unlike other waves, which travel at a particular speed relative to the medium of travel, light travels the same speed for *every* observer—and there is no medium. Thus, for light there are only two speeds—the speed of light, and the *relative* speed of the source of light and the observer. For other kinds of waves, there are three speeds: the speed of the observer relative to the medium, the speed of the source relative to the medium, and the speed of the wave relative to the medium.

But it is more complex because, as speeds approach the speed of light, space and time are interconnected in a manner that somewhat goes against our intuition. This goes to the heart of Einstein's relativity, and we discuss these issues in more detail in Chapter 5. But at relatively low speeds—low, that is, compared to the enormous speed of light—the Doppler effect yields a simple result if one expresses it in terms of wavelength rather than frequency.

If the source and observer are moving relative to each other at some particular percentage of the speed of light, then the observer records a wavelength shifted by that same percentage compared to the laboratory "rest" wavelength. See Equation (4.2):

$$\frac{\Delta\lambda}{\lambda} = \frac{v}{c}, \tag{4.2}$$

where $\Delta\lambda$ is wavelength measured by the observer minus the wavelength emitted by the source, and v is the relative speed between the source and observer. The shift is toward longer wavelengths (called a *redshift*) if the source and observer are moving away from each other, but toward shorter wavelengths (called a *blueshift*) if they are approaching each other.

This Doppler effect for light is of the utmost importance in astronomy; it is how most motion in the universe is directly measured from Earth. Most objects in the universe are far too distant to directly observe a change in direction over the course of a human lifetime. Although the Doppler effect only allows us to measure the part of the motion that is toward or away from Earth, it can be employed no matter how distant the object.

4.6   REFERENCES

Roger Penrose. *The Road to Reality: A Complete Guide to the Laws of the Universe.* Vintage Books, 2004. 67

Barbara Ryden. *Introduction to Cosmology.* Cambridge University Press, 2017. 66

Robert G. Sachs. *The Physics of Time Reversal.* University of Chicago Press, 1987. DOI: 10.1119/1.15494 67

CHAPTER 5

The Present

5.1   SPACE, TIME, AND SPACETIME

In 1905 Albert Einstein published the paper, "On the Electrodynamics of Moving Bodies" and therein described what is now known as *special relativity*. Special relativity resolved conflicts between Newton's laws of motion and the wildly successful theory of electricity and magnetism formalized years earlier by James Clerk Maxwell (building on much earlier work by many others).

Maxwell's equations demonstrated that light is an electromagnetic wave. But the calculated speed of that wave, although in good agreement with experiment, has something strange about it. There is no hint in Maxwell's equations just what this speed is relative to. This is odd indeed, for one of the cornerstones of Newtonian physics is Galileo's principal (also called Newton's first law) that space is relative. And since time for Newton is absolute—it ticks along the same for everyone, once we agree upon a particular standard of measurement—then it follows that velocities must also be meaningful only if measured relative to some agreed upon, but arbitrary, frame of reference.

A velocity is, essentially, space over time. If space is relative and time is absolute, then velocities must be relative. To say that I am traveling at 30 km/s is meaningless, unless I add to that information the detail that it is 30 km/s relative to the pavement of Interstate Route 41. But Maxwell's equation seem to imply that electromagnetic waves travel at a speed that is in some sense absolute.

This discrepancy tied many physicists up in knots at the end of the 19th century, taking them down some dead-end streets. *Einstein solved this problem by dropping the assumption, explicitly made by Newton, that time is absolute.* This freed Einstein to find a way such that there can be *one* speed that is absolute, while all other speeds are relative. That single absolute speed is the speed of light.

It turns out that Maxwell's equations of electricity and magnetism implicitly include this same assumption—but it is hidden, and it took Einstein's analysis to bring this fact to the surface. And so special relativity is in agreement with Maxwell's equations, but in conflict with Newton's laws. Einstein's special relativity of 1905 can be seen as an early part of a revolution in our understanding of physical law. It, along with quantum physics and the GR described in Chapter 11, Section 11.2, are now called *modern physics*, while Newton's laws, Maxwell's equations, and other topics are referred to as *classical physics*.

In Einstein's special relativity, speeds much less than the speed of light—essentially all speeds we experience in our everyday lives—work in very much the same way as they do in Newton's laws, to within the precision of typical measurements. But at relative speeds approaching light speed, the differences are profound. I describe here three important consequences of special relativity, all of which are well verified by experiment, that relate specifically to space and time.

5.1.1   TIME DILATION

The time that passes between any two events depends upon the relative motion of the observer. If I snap my fingers, and then I snap them again, a certain amount of time passes between those two events *as measured by me*. This is called the *proper time interval between the events*:

> The *proper time interval*, t_0, between two events is the time interval as measured in the frame of reference for which the two events happen at the same location in space.

But the interval of time between those same two events is *longer*, as measured in a frame of reference for which the two events take place at different positions in space. And so if you and I are moving relative to each other, *my* two finger snaps occur in the same place for me, but in different places for you—as we move past each other. You measure a different, longer time interval, t, between my finger snaps, given by the time dilation formula:

$$t \;=\; \gamma\, t_0 \tag{5.1}$$

$$\text{where} \quad \gamma \;\equiv\; \frac{1}{\sqrt{1 - \left(\frac{v}{c}\right)^2}}. \tag{5.2}$$

The difference between t, as measured by you, and t_0, as measured by me, is minuscule for our ordinary experience, where relative speeds, v, are much less than the speed of light, c. Table 5.1 shows the time dilation factor, γ (the Greek letter gamma), for different speeds. Notice that it is not really the relative speed v that is important, but rather v/c—how that speed compares to the speed of light.

The relativity factor γ quantifies the magnitude of most relativistic effects. Whenever γ is close to 1, relativistic effects such as time dilation are small, and Newton's laws give answers that, although technically incorrect, are in practice nearly indistinguishable from the correct answer from special relativity. The first three entries in Table 5.1 are so close to unity that many decimal places are needed to show the difference. Even at the enormous speed (from a human perspective) of 10 million meters per second—10,000 times faster than the fastest jet aircraft— γ differs from unity by only 0.056%. And although 99.9% and 99.99999% the speed of light are numerically nearly the same speed, their relativity effects are different by a factor of 100 compared to each other.

5.1.2   LENGTH CONTRACTION

If the time interval between events is not absolute, but instead depends on the relative motion of the observer, then so too must the *distance* traveled between events in spacetime. Let us imagine

Table 5.1: The relativity factor, γ, for different relative speeds, v, compared to the speed of light, c. γ is immeasurably close to unity at speeds much less than the speed of light, and increases without bound as v approaches c.

$v(\text{ms}^{-1})$	v/c	γ
0	0	1
1	3.34×10^{-9}	1.0000000000000000056
1,000	3.34×10^{-6}	1.0000000000056
10,000	0.000334	1.00000000056
0.1000×10^{8}	0.0334	1.00056
2.0000×10^{8}	0.667	1.342
2.9000×10^{8}	0.967	3.945
2.99492666×10^{8}	0.999	22.37
2.99792428×10^{8}	0.9999999	2,236
$\rightarrow c$	$\rightarrow 1$	$\rightarrow \infty$

that I travel to a star that is 2,236 light years from Earth. If, relative to Earth, I make my journey at a speed that is 99.99999% the speed of light, according to Table 5.1, more time will pass on Earth than passes for me by a factor of $\gamma = 2236$. Since, from Earth's point of view, I am traveling at very nearly the speed of light, it would take me about 2,236 years to make the 2,236-light-year trip.

But Table 5.1 indicates that for me much less time would pass—by a factor of 2,236. *And so I would experience the passage of only a single year of time on my trip to this distant star.* Since the speed of light is absolute, and I am traveling at only a tiny fraction less than the speed of light, there is only one conclusion. The *distance* I traveled must be far less—by that same factor of $\gamma = 2,236$.

And so, from Earth's point of view, I travel at very nearly the speed of light, and it takes me 2,236 years to complete the 2,236 light-year distant trip. But because of time dilation I am only one year older when I arrive. As seen from my point of view, as soon as I reach the speed of $v = 0.99999999\,c$, the distance to my destination contracts to only $1/\gamma$ of 2,236 light years— only a single light year. And so, since I am traveling at very nearly the speed of light, it takes me only a single year to make the trip.

And so both points of view agree on the facts: I am one year older when I arrive, but 2,236 years have passed for Earth.

This shortening of lengths along the direction of relative travel is called *length contraction*.[1] We define a *proper length* as follows.

---

[1]It is also sometimes called *Lorentz contraction*, after the physicist H. A. Lorentz, who's work in 1895 was a crucial influence on Einstein's paper of 1905.

The *proper length*, L_0, of an object is its length measured in a frame of reference where it is at rest.

The length of the same object as measured in a frame of reference moving relative to it at speed v, is then given by:

$$L = L_0/\gamma. \tag{5.3}$$

And so we could, in theory, live long enough to travel anywhere in the universe. Simply reach a great enough relative speed such that the distance to travel contracts by an enormous factor γ. It then takes a proportionally shorter amount of time to travel that much-smaller distance. Traveling the universe in your imaginary relativistic spaceship, you would think of things differently. Instead of using rocket fuel to increase your speed, you use it to increase γ, and thus decrease the distance you must travel. Notice from Table 5.1 that the speeds for $\gamma = 22.36$ and $\gamma = 2,236$, for example, are almost the same! But at $\gamma = 2,236$ you have less than 1/100 the distance to travel, to reach the end of your journey, than if you traveled at $\gamma = 22.36$ instead.

The catch is *energy*. To accelerate even a 1-kg mass up to a speed of $\gamma = 2,236$ requires 2.01×10^{20} J of energy—equivalent to more than 48 billion tons of chemical explosives.

5.1.3 THE ADDITION OF VELOCITIES

According to Newton's laws of motion, all velocities are relative to each other in a simple way. If the water in the river flows southward at a speed of 5 m/s relative to the shore, and Valeria and I paddle our canoe southward at 7 m/s *relative to the water*, then we will move southward at a speed of $7 + 5 = 12$ m/s relative to the shore. On our way back on the other hand, as we paddle northward against the current, we will move relative to the shore at a northward speed of only $7 - 5 = 2$ m/s.

Clearly, this simple rule for the combination of velocities cannot hold at speeds near the speed of light. For if it did, then the speed of light would be relative, not absolute. Imagine that I run toward you at 5 m/s, while shining a flashlight ahead of me and toward you. If velocities combined in the simple way proscribed by Newton's laws, you would measure the light from my flashlight coming toward you not at speed $v = c$, but rather at $v = c + 5$ m/s.

Special relativity tells us that velocities combine in a more complex way. Let v and u' represent the two velocities to be combined. And so, for example, v could represent the velocity of the water relative the shore, while u' could represent the velocity of our canoe relative to the water. The velocity, u, of the canoe relative to the shore is then given by:

$$u = \frac{v + u'}{1 + (vu'/c^2)}. \tag{5.4}$$

We can notice something right away about Equation (5.4). If both v and u' are much less than the speed of light, then the denominator is very nearly 1.0. And so at relative speeds much less than c, special relativity gives very nearly the same result as Newton's laws; the velocities simply add.[2]

But at speeds close to the speed of light, the answer is very different. Let us imagine that I run toward you with my flashlight at a speed, v, of 99% the speed of light. And so we can let $v = 0.99c$. The light leaves my flashlight, toward you, at speed $u' = c$. And so what speed, u, do you measure for the light coming toward you from my flashlight? From Equation (5.4) we have:

$$u = \frac{0.99c + c}{1 + (0.99c \cdot c / c^2)} \tag{5.5}$$

$$u = \frac{1.99c}{1.99} \tag{5.6}$$

$$u = c. \tag{5.7}$$

You measure the same speed, c, for the light, even though its source is moving toward you at nearly the speed of light.

5.1.4   MINKOWSKI SPACETIME

Shortly after Einstein's publication of his papers on special relativity, the mathematician Hermann Minkowski described a novel interpretation. He showed that Einstein's equations of special relativity can be given a simple *geometrical* interpretation. If we imagine a four-dimensional *spacetime*, with three spacial dimensions plus the dimension of time, then much of the consequences of special relativity arise in a natural way.

But how can we combine time with space? We have already seen how; we need only express our spacial coordinates of x, y, and z in units of light-seconds in order to put all four axes on the same par. Equivalently, we can simply multiply time by the speed of light, c, to make a time axis that has units of length.

With a four-dimensional spacetime, the consequences of special relativity arise naturally out of what is very nearly ordinary Euclidean geometry, but in four dimensions instead of three. There is a difference however. An important way to describe a particular geometry is with a *metric*—the mathematical rule for calculating distances within that geometry. In ordinary Euclidean geometry, the distance, d, between two points in two dimensions is given by:

$$d^2 = \Delta x^2 + \Delta y^2. \tag{5.8}$$

Here Δx and Δy are, respectively, the difference in the x coordinates and the difference in the y coordinates. This is simply the Pythagorean theorem; the two coordinates x and y form the legs of a right triangle, the hypotenuse of which is our distance d. In three dimensions, with the additional coordinate of z, the distance formula is just as simple:

$$d^2 = \Delta x^2 + \Delta y^2 + \Delta z^2. \tag{5.9}$$

[2]If they are in opposite directions then either v or u' is negative, and the speeds subtract.

And so for straightforward Euclidean geometry, it is clear how to add a fourth dimension of time; simply add another term of $c^2 \Delta t^2$. But what Minkowski showed is that the consequences of special relativity follow from a geometry that is *almost* straightforward four-dimensional Euclidean geometry. The distance formula, however, is slightly different, with a negative sign distinguishing the time coordinate from the space coordinates:

$$S^2 \; = \; d^2 - c^2 \Delta t^2 \qquad\qquad (5.10)$$
$$S^2 \; = \; \Delta x^2 + \Delta y^2 + \Delta z^2 - c^2 \Delta t^2, \qquad\qquad (5.11)$$

where S^2 is called the *spacetime interval*,[3] and it is an example of an *invariant*—a quantity that is the same regardless of the frame of reference. In Newton's laws, both the time between events and the lengths of objects are invariant. But special relativity shows us that this only seems to be true at speeds much less than the speed of light. At speeds near the speed of light, both lengths and intervals of time are relative—this is the meaning of the relativistic phenomena of length contraction and time dilation. *But the spacetime interval, S^2, is the same for all observers.*

It is easy to show that both time dilation and length contraction follow directly from the invariance of this spacetime interval. And so Minkowski's geometrical interpretation of Einstein's special relativity has the advantage of mathematical simplicity and elegance. But there is more to it than that, as Einstein recognized. He used Minkowski's idea of a geometrical interpretation of dynamics as the starting point for *general relativity*, his theory of gravity. We take up this topic in Chapter 11, Section 11.2.

5.2   RIGHT NOW AND RIGHT HERE

The concept of look-back time (Section 4.3) means that our view of the universe necessarily connects space to time. To see is to look back in time, and the further away one looks, the more into the distant past one peers. "A long time ago" really is "in a galaxy far, far away." But Einstein's relativity tells us that this is not simply a consequence of our particular vantage point from Earth. There are real, physical consequences to the fact that space and time are ultimately connected by the speed of light. Since no influence can travel faster than light, spacetime is intertwined with the very relation between physical causes and their effects.

5.2.1   SIMULTANEITY AND THE MEANING OF LOCAL

Because we look back in time as we look out into space, we see the present only in the near-by. It makes human sense to want to know what is happening in some distant galaxy "right now," but we must be cautious about such notions. One lesson of Einstein's special relativity is that time is *not* absolute. And so why should we expect there to be a universal "present" that applies equally both to myself and some alien space-rodent in a galaxy millions of light years distant?

---

[3]Some authors, following Minkowski's original paper, assign the negative signs to the space coordinates and the positive sign to the time coordinate. The mathematical formalism is slightly different, but the physical consequences are the same.

An *event* is a particular point in space and time. We use the term *simultaneity* to refer to different events that happen at the same time. For Isaac Newton, any two events are either simultaneous or they are not. And if *Event B* happened before *Event A*, then that is a fact upon which all agree. But with special relativity, simultaneity is relative. In fact, even the *ordering* of two events may depend upon the relative motion of the observers.

When referring to two events in space and time, it is useful to distinguish between two possibilities.

1. If it is possible for a particle, moving at less than the speed of light, to travel from Event A to Event B, then the two events are said to be separated by a *timelike* interval. If the interval between two events is timelike from the point of view of any one inertial frame of reference, then it will also be timelike from the point of view of every other inertial frame of reference.

2. If it is *not* possible even for a light signal to travel from Event A to Event B, then the two events are said to be separated by a *spacelike* interval. If the interval between two events is spacelike from the point of view of any one inertial frame of reference, then it will also be spacelike from the point of view of every other inertial frame of reference.

And so a spacelike interval is such that the distance, d, measured in light years between the two events is greater than the number of years, Δt, that a light signal would require to travel that distance. A timelike interval is one for which the opposite is true. It should be clear that in terms of our spacetime interval, as defined in Equation (5.11), S^2 is a positive number for a spacelike interval and a negative number for a timelike interval. We can also talk about the borderline case between these two intervals, whereby it requires exactly the speed of light to travel from one event to the other. Such an interval is called *lightlike*.

If S^2 can be negative, what about S itself? Wouldn't that involve the square root of a negative number, resulting in so-called *imaginary numbers*? Indeed it would, and therein lies much of the power of Minkowski's geometric interpretation of special relativity. The mathematics involving imaginary numbers is called *complex analysis*, and it lends itself to a richer theoretical structure than the ordinary *real numbers* associated with Newton's laws. And although imaginary numbers and complex analysis can be used in the mathematics of special relativity [Penrose, 2004, p. 414], it always works out that calculated quantities that can be measured are ordinary real numbers.

The distinction between spacelike and timelike intervals is crucial, because nothing can travel faster than light. And so *if two events are separated by a spacelike interval, then it is impossible for either event to have caused the other*. Events separated by a timelike interval, on the other hand, may be *causally connected*; it is physically possible that one of the events might have caused the other.

In special relativity, *simultaneity is relative*. If in some particular frame of reference two events happen at different places but at the same time, then there will always be other frames of reference for which one of the events happened before the other. But notice that if this is true, then the two events must be separated by a spacelike interval; to say that the events are simultaneous in some particular reference frame is to say that although they are separated by space, there is zero time between the events. And so not even a light beam could connect them.

And so we are left with the notion that Event A might precede Event B in one frame of reference, while in another reference frame the two events happened simultaneously, and in yet a third frame of reference Event B happened before A. But whenever this situation arises, the two events are always separated by a *spacelike* interval, and so they cannot be causally connected. Whenever two events are instead separated by a *timelike* interval, all observers will agree upon which event happened first. Thus, special relativity maintains *the principle of causality*: physical causes always happen *before* their effects.

5.2.2 TIME VARIATIONS AND SIZE

There are practical applications to the notion in special relativity that events that are causally connected are always related by timelike intervals in spacetime. Many astronomical objects, so distant that they appear as only points of light, are observed to vary in brightness on short time scales. The relativistic principle of causality suggests that if, for example, an astronomical body substantially increases and decreases its brightness over a period of only 12 hr, *then it must have a physical size of 12 light-hours or less*. For how else could the different parts "know" to increase or decrease their brightness in concert with each other, if the different parts of the object can have no causal connection to each other?

This type of analysis has been especially important in the study of many type of *active galactic nuclei (AGN)*, extremely powerful sources of energy seen at the centers of many distant galaxies. Short timescale variations in brightness are often observed in many of these objects, and this sets limits to the volume of space from which this power is emitted [Wagner and Witzel, 1995]. The conclusion in many cases is that AGN are powered by very compact sources of energy, and so theories of the source of power for these objects must conform to these size limits.

The prevailing theory of the "central engine" of AGN is that matter falling into a supermassive black hole is the ultimate source of power for AGN, and restrictions set by light variations and the principle of causality are important pieces of the circumstantial evidence for this conclusion [Carroll and Ostlie, 2017, pp. 1109–1110].

5.3 COSMOLOGY AND THE COSMOLOGICAL PRINCIPAL

The study of the universe as a whole, on the large scale, and its history over time, is called *cosmology*. In its modern form, we recognize that the properties of the universe *on the large scale* change over time. But because of the relativistic connection between space and time, we describe these overall properties in terms of what would be observed by a *comoving observer*—a hypothetical person riding along with the evolution of the universe, making observations as time passes.

A crucial assumption is the *cosmological principle*:

The Universe is assumed to be—on the largest of scales—*homogeneous* and *isotropic*. To say the universe is homogeneous means that it is the same everywhere. To say that is isotropic is to say that it is the same in every direction.

The cosmological principal is a starting point for doing cosmology. If the consequences of assuming the cosmological principal to be true turn out to be in serious conflict with observations, then we will have to drop those assumptions. But so far, and perhaps surprisingly, there has been no compelling evidence that the cosmological principal is incorrect.

5.4 COSMOLOGICAL PARAMETERS

A particular *cosmological model* applies the laws of physics, as best we understand them, to the universe as a whole, following along with the changes that would occur. These physical laws alone are not enough, one must still choose the value of various *parameters*—numerical quantities that have no known particular theoretical value, but that greatly affect the outcome of the cosmological model. Some of these parameters can be directly measured with astronomical observations. Others are adjusted such that the model achieves the best agreement with actual observations of the universe. And so a particular cosmological model is a combination of the physical laws and a particular set of cosmological assumptions, with parameters adjusted so as to accommodate the best agreement with many different observations.

In Table 5.2 I list some modern estimates of several of these parameters, and I describe each of them below. All of the parameters I list here are meant to describe the *present* state of the universe. The cosmological model then assumes these values, while applying the laws of physics to describe the universe in both the past and the future. In the sections that follow, I describe each of these parameters in turn.

5.4.1 THE HUBBLE PARAMETER (H_0) AND HUBBLE'S LAW

All galaxies, except for the few that are relatively close to us, show a *redshift* in their spectrum. The spectrum of most galaxies show absorption lines (some also show emission lines), and these lines are inevitably all shifted, by the same percentage, toward *longer* wavelengths.

This is just what one would expect from the Doppler effect, described in Section 4.5. And so it would seem to mean that all galaxies are moving *away* from us. The redshift, z, is defined in terms of the shift in wavelength, λ, as follows:

$$z = \frac{\Delta\lambda}{\lambda} = \frac{\lambda_{\text{observed}} - \lambda_0}{\lambda_0}, \tag{5.12}$$

where λ_0 is the known wavelength as measured in the laboratory—presumably the actual wavelength emitted by the galaxy. And so z represents the fractional shift in the observed wavelength.

Table 5.2: Cosmological parameters that represent the present state of the universe. The measured values are taken from Planck Collaboration et al. [2016, Table 4, column 3].

Name	Symbol	Value
Hubble parameter	H_0	67.9 km s^{-1} Mpc^{-1}
Current Age of Universe	t_0	13.8×10^9 years
Baryon Density	Ω_b	0.048
Dark Matter Density	Ω_c	0.26
Dark Energy Density	Ω_λ	0.69
Current CMB Temperature	T	2.73 K
Current average hydrogen abundance	X	0.74
Current average helium abundance	Y	0.23
Current average metallicity	Z	0.02

If we interpret the redshift as arising from the Doppler effect, then we can combine Equations (5.12) and (4.2) to calculate the speed, v, at which a given galaxy with measured redshift is receding from us:

$$v = zc. \tag{5.13}$$

In the years between 1912 and 1929, Vesto Slipher and Edwin Hubble measured the redshifts of many galaxies. Combined with estimates of the *distances* to these galaxies, a famous relation became apparent, now known as *Hubble's Law*:

> For galaxies that are outside of our local neighborhood, but still less than several billion light years away, *the redshift of a galaxy is directly proportional to its distance.*

And so combining this observation with Equation (5.13), we can express Hubble's law with the following simple equation:

$$v = H_0 d, \tag{5.14}$$

where d is the distance to the galaxy and H_0 is the *Hubble parameter* or *Hubble constant*.

From Equation (5.14) it is clear that the Hubble parameter has dimensions of speed divided by length, and it is usually expressed in km/s of velocity per megaparsec of distance. These odd units are chosen to be in line with typical measurements of both v and d for galaxies. In order to determine the Hubble parameter, one must measure both the redshifts and distances of many galaxies. These two values can then be plotted on a graph with v (in km/s) on the vertical axis and d (in Mpc) on the horizontal axis; H_0 is then the simply the slope of this graph. The redshift is the easy part; it is the distance that is difficult to measure.

A modern example of this *velocity-distance relation* or *Hubble diagram* can be seen in Figure 5.1. Note that the galaxies in the Virgo cluster show a large spread of velocities for a given

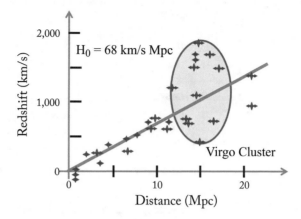

Figure 5.1: The Hubble diagram for relatively nearby galaxies. The slope gives the Hubble constant. (Graphic by Brews ohare, CC BY-SA 3.0.)

distance. This is because their motions are not only due to Hubble's law, but also simply because they are in a large galaxy cluster, and they move according to the gravitational forces they feel for each other. This brings up an important point. Hubble's law applies to motions on the large scale—over small distances galaxies move according to the local gravitational forces they feel for each other.

Equation (5.14) is commonly used to describe Hubble's law, and it is the way Hubble presented it in 1929, before the birth of modern cosmology [see Ryden, 2017, p. 12ff]. But the modern cosmologist is more careful; it is not the velocity of the galaxy that is directly measured, but rather its *redshift*. This redshift is only a velocity if we assume that it is directly caused by a straightforward Doppler effect. At relatively small distances, this assumption works, but in general it is incorrect—the redshifts of galaxies are instead caused by the expansion of space itself.

The value of the Hubble parameter can be measured or inferred by many different methods. See Figure 5.2 for the recent history of measurements. The most precise measurements (the smallest error bars) are in reasonable agreement with each other, for a Hubble parameter of about 70 km/s per Mpc. Values from many different methods vary between about 67–74 km/s per Mpc.

5.4.2 THE EXPANSION OF THE UNIVERSE AND THE BIG BANG

The simplest explanation for Hubble's law, in accordance with the cosmological principal, is that the universe is undergoing a *uniform expansion*. That is to say, over a given period of time, *all distances in the universe increase by the same percentage*.

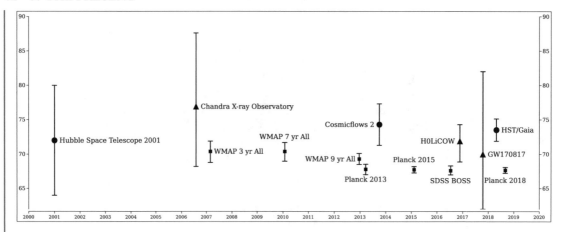

Figure 5.2: Different measurements of the Hubble constant over recent years. The vertical bars represent the uncertainty in the measurements. The values are in fair agreement with each other, even though they use a variety of independent methods. (Graphic by Kintpuash—Own work, CC0.)

For a good analogy, make some raisin bread. Knead the dough and let it rise. As the dough expands, it carries the raisins with it, and *every* raisin would see all of the other raisins getting farther and farther away. The raisins themselves do not expand; they are simply carried along by the expanding dough-space. Furthermore, more distant raisins recede *faster* because the expansion is proportional. If, for example, the rising dough doubles in size in three hours, then a raisin initially 1 cm away will now be 2 cm distant—and so it would have receded at a rate of $2 - 1 = 1$ cm per three hours, or $0.33\,\mathrm{cm\,h^{-1}}$. By the same logic, however, a raisin that is 10 cm distant initially would have receded at the proportionally larger speed of $3.3\,\mathrm{cm\,h^{-1}}$. And so tiny astronomers living on the raisin would measure a raisin bread version of Hubble's law with their tiny telescopes.[4]

Note that this uniform expansion preserves the cosmological principal. Tiny astronomers on *any* rasin would observe the same thing; the raisin bread universe is *homogenous*. And it only matters how far apart two raisins are, not their directions, so our expanding-dough universe would also be *isotropic*. The model of a uniformly expanding universe is the simplest explanation for the observation of Hubble's law, if we assume the cosmological principal is correct.

A flaw in our raisin bread analogy is that the dough has an edge, while the universe does not. And space and time are not connected together in this raisin bread universe, in the profound and strange way that they are in the real universe.

Given that the universe is expanding, a simple explanation is that it means just what it seems to; everything is getting farther apart as time passes. And so in the past everything was

---

[4]Presumably their telescopes would operate at wavelengths that could see through dough.

closer together. Furthermore, there would have been a time when everything was all in the same place at once. This is the basic idea behind the *Hot Big Bang*:

> The universe was once infinitely hot and dense, and it has been expanding and cooling ever since. Thus, there was a $t = 0$ when the density of the universe was infinite. That is, *the universe has a finite age*.

And so our expanding Big Bang universe is the same in every direction, and the same everywhere. But it is not the same every*when*.

5.4.3 THE CURRENT AGE OF THE UNIVERSE (t_0)

Although the Hubble parameter is usually expressed in $\mathrm{km\,s^{-1}\,Mpc^{-1}}$, it has units of inverse time hidden within it; to see this, simply note that it is a length/time per length. We can thus express the reciprocal of the Hubble parameter, $1/H_0$, in years. This is called the Hubble time, t_H, and converting the units in a convenient way, it is given as follows:

$$t_H \text{ (billions of years)} = \frac{976}{H_0 \,(\mathrm{km\,s^{-1}\,Mpc^{-1}})}. \tag{5.15}$$

For $H_0 = 67.7$, this gives a Hubble time of $t_H = 14.4$ billion years.

The Hubble time represents the age of the universe if it were to expand at a uniform rate throughout its history. In the absence of additional information, this would be a reasonable first guess for the actual age of the universe. But we do not expect this to be the actual age, for the simple reason that we have good reasons to believe the universe did *not* expand at a uniform rate. A full cosmological model is required to make a good estimate of the current age of the universe. We use t_0 to denote this current age, and our best value as of 2019 is $t_0 = 13.8$ billion years, just a bit less than the Hubble time.

5.4.4 THE BARYON DENSITY (Ω_b)

The so-called baryon density refers to the average density of "ordinary" matter in the universe. This is matter made ultimately of familiar particles such as protons, neutrons and electrons—the basic constituents of atoms. It is a little bit of a misnomer, because technically electrons are not baryons; they belong to a different category of particles called *leptons*. But protons and neutrons have nearly 2000 times the mass of electrons, so it is perhaps not too much of an oversight to use a name that, technically, should not include electrons. In practice, "baryon density" simply means the density of what is usually called "matter."

As is the case with Hubble's law, we mean the density on average over a large range of distances that encompasses not only galaxies or even clusters of galaxies, but superclusters as well. The current estimate is that the baryon density makes up only about 4.8% of the universe. And so even though the baryon density represents the kind of matter that *matters* most to us—it is related to all that we see and touch—it makes up only 4.8% of the universe. What is the rest? See Sections 5.4.5 and 5.4.6.

Table 5.3: The three constituents of the universe, along with our degree of understanding of each

Component	Percentage of the Universe	Amount of our Knowledge
Normal matter	5%	A lot
Dark matter	26%	Very little
Dark energy	69%	Almost nothing

5.4.5   THE DARK MATTER DENSITY (Ω_c)

There is evidence, accumulated over many years, that much of the gravity in the universe—and thus much of the mass—produces no light, nor is it otherwise affected by normal matter. We call this unseen matter *dark matter*. We can detect dark matter even though it is invisible; it still has gravity, and so it affects the *motions* of ordinary baryonic matter that we *can* see.

The most compelling idea currently is that dark matter is some type of hitherto undetected subatomic particle that has properties quite unlike the familiar types that make up atoms. It is proposed that these *weakly-interacting massive particles* (WIMPS) have mass—and thus gravity. But they do not produce or otherwise interact with light. Nor are they affected by ordinary matter, except through gravity.

The current estimates are that dark matter makes up about 26% of the universe—and so there is roughly five times as much dark matter as ordinary (baryonic) matter.

5.4.6   THE DARK ENERGY DENSITY (Ω_λ)

Beginning in the late 1990s, evidence accrued that suggests the expansion rate of the universe is currently speeding up. This was both a surprising and transforming discovery; much observational data suddenly fit together better than ever before into one complete whole. But on the other hand, the *physical* explanation for this *accelerating expansion* is unknown. The more-or-less agreed upon name for this observed accelerating effect is *dark energy*.

Observations suggest that dark energy is the *biggest* component of the universe, accounting for 69%, as opposed to 26% for dark matter and only 5% for normal matter. And so we are left with the odd circumstance illustrated by Table 5.3. What seems to be most prominent in the universe is what we understand least.

5.4.7   THE COSMIC MICROWAVE BACKGROUND TEMPERATURE (T)

The cosmic microwave background (CMB) is a dim glow of microwaves seen in every direction. Its most compelling explanation is that it is radiation left over from when the universe was extremely hot and dense. We will have much more to say about it, but it can be described very precisely by its *temperature*. Figure 5.3 shows the precisely measured energy distribution of the

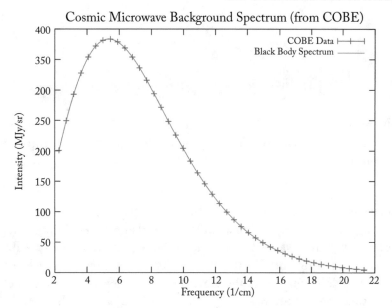

Figure 5.3: The observed spectrum of the cosmic microwave background together with a theoretical curve for a temperature of 2.73 K. The error bars for the observed data are too small to show on the diagram. (Graphic by Quantum Doughnut, Public Domain.)

CMB, along with a theoretical curve with a temperature of 2.73 K, less than three degrees above absolute zero.

5.4.8   THE CHEMICAL COMPOSITION OF THE UNIVERSE (XYZ)

The light we receive from stars, galaxies, and clouds of gas can tell us many things about the matter from which it was emitted. In particular, an analysis of the *spectrum* of the light often allows us to determine—sometimes to high precision—the chemical composition of the gases that emitted the light. We explore how that works in more detail in Part IV of *The Big Picture*, but let us here look at the results. What kinds of atoms are stars, galaxies, and clouds of gas and dust typically made of?

The answer turns out to be the same over and over again, no matter where we look. Overall, the universe seems to be made everywhere of very roughly these percentages of atoms (more precise average values can be seen in Table 5.2):

- 3/4 Hydrogen,
- 1/4 Helium, and
- 0–2% Everything else put together.

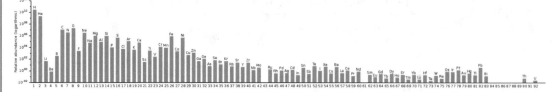

Figure 5.4: The relative abundances of the chemical elements in the solar system, portrayed on a logarithmic scale. The horizontal axis is atomic number (number of protons in the atom). (Graphic by Swift, CC0.)

This pattern repeats itself over and over. There are variations; sometimes the helium abundance is a little higher and the hydrogen abundance a little lower. And the small percentage of "everything else put together" varies between nearly zero and about 2% or so. But even in this case, the similarities are intriguing. For those other elements besides hydrogen and helium tend to vary from place to place with nearly the same abundances relative to each other.

Since it is tiresome to keep repeating "everything else put together," or "everything but hydrogen and helium," a word is needed. For historical reasons, astronomers use the rather unfortunate term *metals* to refer to all of the elements, taken collectively, besides hydrogen and helium. This would make a chemist cringe; the most common of the "metals" astronomers refer to are carbon (C), nitrogen (N), and oxygen (O). These three are about as non-metallic as an element can be. To add insult to injury, we refer to an astronomical object's percentage abundance of metals with the totally-made-up-by-astronomers word *metallicity*.

There are intriguing patterns in the abundances of "metals":

- Roughly speaking, the more massive an atom is, the *less* abundant it is in the universe.
- Atoms with an odd number of protons are slightly less abundant than atoms with an even number of protons.

We will have much to say about these facts later, in Section 9.11. But here we note that these patterns hold for both relatively large and small overall abundances of the metals.

See Figure 5.4 for an illustration of the abundances of atoms in the solar system. This is just a tiny dot in the universe, but most of the matter in the solar system is the Sun; the planets add very little to the total. And so the relative abundances show in Figure 5.4 are very close to those for the Sun itself. And the Sun is a rather typical star—albeit a star with comparatively *high* metallicity. Note the overall pattern of decreasing abundance with number of protons, and the zig-zag alternating pattern of high and low abundance from even-numbered to odd-numbered elements.

Figure 5.4 shows the relative abundances on a *logarithmic scale*; looking at the numbers on the vertical axis, we can see that carbon, for example, is thousands of times less abundant than hydrogen. If we were to represent the abundances with an ordinary linear scale, only hydrogen and helium would show on the graph.

5.5   REFERENCES

Bradley W. Carroll and Dale A. Ostlie. *An Introduction to Modern Astrophysics*, 2nd ed., Cambridge University Press, 2017. DOI: 10.1017/9781108380980 78

Roger Penrose. *The Road to Reality: A Complete Guide to the Laws of the Universe*. Vintage Books, 2004. 77

Planck Collaboration, P. A. R. Ade, N. Aghanim, M. Arnaud, M. Ashdown, J. Aumont, C. Baccigalupi, A. J. Banday, R. B. Barreiro, J. G. Bartlett, N. Bartolo, E. Battaner, R. Battye, K. Benabed, A. Benoît, A. Benoit-Lévy, J. P. Bernard, M. Bersanelli, P. Bielewicz, J. J. Bock, A. Bonaldi, L. Bonavera, J. R. Bond, J. Borrill, F. R. Bouchet, F. Boulanger, M. Bucher, C. Burigana, R. C. Butler, E. Calabrese, J. F. Cardoso, A. Catalano, A. Challinor, A. Chamballu, R. R. Chary, H. C. Chiang, J. Chluba, P. R. Christensen, S. Church, D. L. Clements, S. Colombi, L. P. L. Colombo, C. Combet, A. Coulais, B. P. Crill, A. Curto, F. Cuttaia, L. Danese, R. D. Davies, R. J. Davis, P. de Bernardis, A. de Rosa, G. de Zotti, J. Delabrouille, F. X. Désert, E. Di Valentino, C. Dickinson, J. M. Diego, K. Dolag, H. Dole, S. Donzelli, O. Doré, M. Douspis, A. Ducout, J. Dunkley, X. Dupac, G. Efstathiou, F. Elsner, T. A. Enßlin, H. K. Eriksen, M. Farhang, J. Fergusson, F. Finelli, O. Forni, M. Frailis, A. A. Fraisse, E. Franceschi, A. Frejsel, S. Galeotta, S. Galli, K. Ganga, C. Gauthier, M. Gerbino, T. Ghosh, M. Giard, Y. Giraud-Héraud, E. Giusarma, E. Gjerløw, J. González-Nuevo, K. M. Górski, S. Gratton, A. Gregorio, A. Gruppuso, J. E. Gudmundsson, J. Hamann, F. K. Hansen, D. Hanson, D. L. Harrison, G. Helou, S. Henrot-Versillé, C. Hernández-Monteagudo, D. Herranz, S. R. Hildebrand t, E. Hivon, M. Hobson, W. A. Holmes, A. Hornstrup, W. Hovest, Z. Huang, K. M. Huffenberger, G. Hurier, A. H. Jaffe, T. R. Jaffe, W. C. Jones, M. Juvela, E. Keihänen, R. Keskitalo, T. S. Kisner, R. Kneissl, J. Knoche, L. Knox, M. Kunz, H. Kurki-Suonio, G. Lagache, A. Lähteenmäki, J. M. Lamarre, A. Lasenby, M. Lattanzi, C. R. Lawrence, J. P. Leahy, R. Leonardi, J. Lesgourgues, F. Levrier, A. Lewis, M. Liguori, P. B. Lilje, M. Linden-Vørnle, M. López-Caniego, P. M. Lubin, J. F. Macías-Pérez, G. Maggio, D. Maino, N. Mandolesi, A. Mangilli, A. Marchini, M. Maris, P. G. Martin, M. Martinelli, E. Martínez-González, S. Masi, S. Matarrese, P. McGehee, P. R. Meinhold, A. Melchiorri, J. B. Melin, L. Mendes, A. Mennella, M. Migliaccio, M. Millea, S. Mitra, M. A. Miville-Deschênes, A. Moneti, L. Montier, G. Morgante, D. Mortlock, A. Moss, D. Munshi, J. A. Murphy, P. Naselsky, F. Nati, P. Natoli, C. B. Netterfield, H. U. Nørgaard-Nielsen, F. Noviello, D. Novikov, I. Novikov, C. A. Oxborrow, F. Paci, L. Pagano, F. Pajot, R. Paladini, D. Paoletti, B. Partridge, F. Pasian, G. Patanchon, T. J. Pearson, O. Perdereau, L. Perotto, F. Perrotta, V. Pettorino, F. Piacentini, M. Piat, E. Pierpaoli, D. Pietrobon, S. Plaszczynski, E. Pointecouteau, G. Polenta, L. Popa, G. W. Pratt, G. Prézeau, S. Prunet, J.

L. Puget, J. P. Rachen, W. T. Reach, R. Rebolo, M. Reinecke, M. Remazeilles, C. Renault, A. Renzi, I. Ristorcelli, G. Rocha, C. Rosset, M. Rossetti, G. Roudier, B. Rouillé d'Orfeuil, M. Rowan-Robinson, J. A. Rubi no-Martín, B. Rusholme, N. Said, V. Salvatelli, L. Salvati, M. Sandri, D. Santos, M. Savelainen, G. Savini, D. Scott, M. D. Seiffert, P. Serra, E. P. S. Shellard, L. D. Spencer, M. Spinelli, V. Stolyarov, R. Stompor, R. Sudiwala, R. Sunyaev, D. Sutton, A. S. Suur-Uski, J. F. Sygnet, J. A. Tauber, L. Terenzi, L. Toffolatti, M. Tomasi, M. Tristram, T. Trombetti, M. Tucci, J. Tuovinen, M. Türler, G. Umana, L. Valenziano, J. Valiviita, F. Van Tent, P. Vielva, F. Villa, L. A. Wade, B. D. Wandelt, I. K. Wehus, M. White, S. D. M. White, A. Wilkinson, D. Yvon, A. Zacchei, and A. Zonca. Planck 2015 results. XIII. Cosmological parameters. *Astronomy and Astrophysics*, 594:A13, September 2016. DOI: 10.1051/0004-6361/201525830 80

Barbara Ryden. *Introduction to Cosmology*. Cambridge University Press, 2017. 81

S. J. Wagner and A. Witzel. Intraday Variability in Quasars and BL Lac Objects. *Annual Review of Astronomy and Astrophysics*, 33:163–198, January 1995. DOI: 10.1146/annurev.aa.33.090195.001115 78

CHAPTER 6

The Past

6.1 MEASURING THE HISTORY OF THE UNIVERSE

The concept of look-back time allows us to record a history of the expansion of the universe. The Hubble parameter gives us the *current* expansion rate of the universe; if we extend the Hubble diagram of Figure 5.1 to vast distances, we sample that expansion rate further and further back in time.

The greater the slope of the Hubble diagram, the faster the universe expands. If, for example, the slope is *greater* for very large distances, it means the universe was expanding *faster* in the distant past. The opposite would be true if the slope turns out to be smaller at large distances. See Figure 6.1 for examples of Hubble diagrams that compare a constant expansion rate to hypothetical universes that were expanding either more slowly or more rapidly in the past.

In practice, cosmologists do not directly plot v vs. d. Instead the directly observed redshift, z, is used instead of a velocity calculated from the Doppler effect formula. Although a simple Doppler effect interpretation of the cosmological redshift is convenient for relatively small distances, it is not really correct when one considers the details of an expanding universe. And so cosmologists use detailed cosmological models to predict the relationship between redshift and some observable quantity that relates to distance. The predictions from the models are then compared directly to the observations.

6.2 THE BEGINNING

If we take the observed expansion of the universe at face value—that it means the universe was once much hotter and denser, this implies there was a beginning to the universe we now observe, a $t = 0$ when the temperature and density was infinite. What came before? One may speculate (and many do), but such speculations are not really part of what we call *big bang cosmology*.

And so we imagine the universe expanding from infinite temperature and density, cooling and becoming less dense with the passage of time. This basic idea has been fleshed out in quite a bit of detail over the past few decades. One could now even go so far as to say that it actually makes a fair amount of sense. The Big Bang is intriguing because it explains, with surprisingly few assumptions, four key overall features of the present universe:

1. The universe is made up of 3/4 Hydrogen and 1/4 Helium, with only traces of everything else put together. Note that this refers only to normal matter.

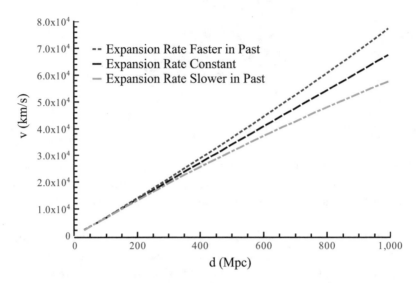

Figure 6.1: The Hubble diagram for different possible universes. The straight-line black curve shows a universe expanding at a constant rate. The red line shows a universe that was expanding faster in the past—and so the curve is more steeply sloped at large distances (greater look-back time). This is what one would expect with ordinary gravity slowing the expansion over time. The green line shows the opposite, an accelerating universe expanding more rapidly in the present than the past.

2. The universe has a finite age. That is, there is a maximum age to the stuff we see in the universe.

3. The universe is expanding in a uniform way, as shown by the observation of Hubble's law.

4. The universe is currently filled with a uniform radiation, at a temperature of 2.77 K—the CMB.

In addition, the Big Bang has recently made a fair amount of progress in connecting smoothly with other theories that try to explain the following.

1. The way that galaxies, galaxy clusters, superclusters, and large-scale structure developed in the universe.

2. Some properties of the mysterious dark matter.

3. The nature of the fundamental laws of physics and how they relate to each other in a unified way.

6.3   A TIME-LINE TO NOW

Following is a condensed timeline of key events in the early history of the universe, summarized in Table 6.1, expressed in terms of what is probably our best current overall cosmological theory, the so-called Λ-CDM model. We will consider the details of the Λ-CDM model more fully in Section 17.1; here we simply describe its implications for the history of the universe.

The variable t refers to how much time has passed since $t = 0$. The variable T stands for temperature, that is, how hot the universe is. The temperature is expressed in the Kelvin scale, which tells how many degrees Centigrade above absolute zero. But first off, we must be clear regarding what we mean by t and T. Who is measuring these variables? In particular, we have already seen that time is relative. And so *who's* time do we mean? The answer is that we describe the universe in terms of a *co-moving observer*—a hypothetical person riding along with the expansion of the universe, describing the conditions as they happen *for them*. And so t represents the time ticking along on *their* clock, but we can generalize this result; because of the cosmological principal, we assume that any other observer would experience the same.

For the first 380,000 years or so, the baryonic (ordinary) matter and electromagnetic radiation (light) in the universe are locked together, and so both are at the same temperature. After this *time of decoupling*, baryonic matter and light go there separate ways, and so the temperature of the universe is less meaningful as a concept for the matter. But it still has some meaning for the radiation, and so that is what T refers to after that time. The redshift, z, is also listed, but it is important to note that for many of the earlier entries we have no way to directly measure such a redshift for any observable object. Notice that for smaller t (longer ago), z is bigger. In the sections that follow, we briefly consider each part of this chronology.

6.3.1   THE PLANCK ERA

Before this time, ending at about $t = 10^{-43}$ s, the universe is too hot and dense for our established physics to apply. To understand what is going on at this high temperature and density, we would need to know how all the forces of nature are connected to each other. The *Standard Model of Particle Physics* explains most of the relations between electricity, magnetism and the forces at work within the nucleus of an atom. But it does not include gravity in a unified way, and it would surely need to in order to describe the conditions at temperatures this high.

6.3.2   INFLATION

At about 10^{-36} s it is thought that the so-called strong force, that holds protons and neutrons together in the nucleus of an atom, separated from the other fundamental forces. This may have caused a brief and sudden, exponential expansion of the universe called *inflation*, an idea first proposed by Alan Guth [1981].

Inflation is an attractive proposal for three reasons. First, it has some theoretical justification from the standpoint of fundamental physics, albeit physics that is at best only partially

Table 6.1: **A chronology of key events in the history of the universe.** The table shows the redshift, time from the beginning of the big bang in both seconds and years, the time in years from the present, and the approximate temperature of the radiation in the universe. (The specific values are approximate, and derived from a particular cosmological model, taken mostly from en.wikipedia.org/wiki/Chronology_of_the_universe.)

Event	z	t (s)	t (yrs)	t_0-t (yrs)	T(K)
Beginning of Big Bang	∞	0	0	13.8×10^9	∞
End of Planck era		1×10^{-43}		13.8×10^9	10^{32}
End of inflation era		1×10^{-32}		13.8×10^9	10^{22}
Protons and neutrons form		1×10^{-6}		13.8×10^9	10^9
Proton-neutron ratio fixed		1.0		13.8×10^9	10^8
End of fusion era		180		13.8×10^9	10^7
Time of decoupling	1,100	1.2×10^{13}	3.8×10^5	13.8×10^9	4,000
Re-ionization begins	10	2×10^{16}	5×10^8	13.3×10^9	60
Formation of Milky Way	7,6	2.4×10^{16}	6.8×10^8	13.1×10^9	40
Formation of Sun	0.423	2.81×10^{17}	9.17×10^9	4.63×10^9	4.1
Formation of Earth	0.418	2.89×10^{17}	9.23×10^9	4.57×10^9	4.1
Dark energy era begins	0.4	3×10^{17}	9.8×10^9	4×10^9	4
The present (t_0)	0	4.33×10^{17}	13.8×10^9	0	2.7

understood. Second, it solves the so-called *horizon problem*. Finally, inflation solves the so-called *flatness problem*. We discuss these issues further in Chapter 17.

6.3.3   THE FORMATION OF PROTONS AND NEUTRONS

After inflation ended at about 10^{-32} s, the universe resumed its previous much-slower (but still rapid!) expansion. At this point it was a sea of fundamental particles: quarks, leptons, and photons. Quarks (the name was pulled by physicist Murray Gell-Mann from James Joyce's novel, Finnegan's Wake) are the basic building blocks of protons and neutrons, which reside at the nuclei of atoms. At these temperatures, however, quarks cannot hold together long enough to form these familiar particles. Leptons are another class of particles, the most familiar of which is the electron. Photons are particles of light.

By about 1 µs the universe had cooled enough that protons and neutrons could exist. But it was still too hot and dense for them to be stable, and so they rapidly changed back and forth into each other. That is, protons changed into neutrons, but neutrons also changed into protons. Both processes would have occurred simultaneously, but in a lopsided way: it is more likely for a neutron to change into a proton than the reverse. This means that there was more of a tendency to make protons than to make neutrons.

As the universe cools further, by around $t = 1\,\text{s}$ it is no longer hot enough for neutrons and protons to change into each other, and so we are now stuck with the results. Computer calculations show that when all was said and done, there should have been about seven protons for every neutron. Since the universe, after this point in time, is too cool to change that situation, it is what we should have now. And so Big Bang cosmology here makes a fundamental prediction regarding the universe today: there should be seven protons for every neutron in the universe. That is, unless parts of the universe later find some way to heat itself back up to these enormous temperatures.

6.3.4   THE FUSION ERA

Immediately after stable neutrons and protons have formed, it is too hot for them to get together to form the nuclei of atoms. But as the temperature drops, protons and neutrons begin to join up and form familiar and stable nuclei. After about 3 min have passed, the universe has cooled to below 10 million Kelvin, and it is thereafter *too cool* for protons and neutrons to rearrange their combinations; we are then stuck with whatever has formed by that time.

This is called the *fusion era* because it is similar to the process of nuclear fusion that occurs in the hot and dense centers of stars such as the Sun. In a sense, during the fusion era the conditions everywhere throughout the universe are not unlike the conditions found in the centers of stars.

It turns out that of all the possible ways for protons and neutrons to hook up with each other, there are two ways that are far more likely than any other.

1. *Single protons*: a proton all by itself is the nucleus of a normal hydrogen atom.

2. *Two protons joined with two neutrons*: this is the nucleus of the most common form of helium.

And so the neutrons and the protons hitch up almost entirely in these two ways. To make one helium nucleus, two neutrons are needed. Since 7 protons were made for every neutron, there must be $7 \times 2 = 14$ protons for these 2 neutrons. Two of them will be needed to complete that helium nucleus, leaving 12 with no neutrons to hitch up with. This means we get one helium nucleus (2 protons with 2 neutrons) for every 12 hydrogen nuclei (single protons). But a helium nucleus weighs four times as much as a hydrogen nucleus, so *by weight* we have the result in Table 6.2: 75% hydrogen, 25% helium.

Since the universe is cooling, after these first 3 min, it is too cool to break apart nuclei or fuse them together. And so we are stuck with a universe made of 3/4 hydrogen and 1/4 helium. Thus, Big Bang cosmology makes another prediction about the universe, and we have already seen that this is roughly the observed composition of the universe today.

Table 6.2: Hydrogen and helium nucleosynthesis

1	Helium	weighs	4
12	Hydrogens	weigh	12
Total		weighs	16
Helium/total is	4/16	=	25%
Hydrogen/total is	12/16	=	75%

But again, if the universe can later find some way to heat parts of itself up to these same temperatures, other nuclei could be formed. I have already hinted that this *does* happen, and stars such as the Sun are the key.

6.3.5   TIME OF DECOUPLING

Before about 380,000 years the universe was too hot for electrons to join up with the hydrogen and helium nuclei to make neutral atoms. But when the universe cooled to about 3000 K, free electrons could join up with hydrogen and helium nuclei to make neutral atoms. This is called the epoch of recombination. It is something of a misnomer; electrons are not *re*-combining with nuclei, they are combining for the first time!

Before this, the universe was opaque. Photons of light interact very strongly with free electrons, so light could not travel very far before scattering off in another direction. When the universe was half a million years old, the free electrons were no longer free; they joined up with hydrogen and helium nuclei to make neutral hydrogen and helium atoms. Thus, there was no longer anything for light to scatter off of, and the universe rather suddenly became transparent.

Before the epoch of recombination, the intense radiation would have kept the protons, neutrons, and electrons at a uniform density; it would have been difficult for a denser clump here, a less-dense clump there to form. After this time, when we talk about the temperature of the universe, we are talking about the temperature of the leftover radiation. Matter, meanwhile, is free to do its own thing, while the radiation continues to cool.

6.3.6   TIME OF RE-IONIZATION: STARS AND GALAXIES

The central idea of Big Bang cosmology is that the universe began extremely hot and dense, and rapidly expanded and cooled. But meanwhile, gravity was working on matter, pulling little bits together to make small parts of the universe hot and dense again. And so the universe—as a whole, on the largest scales—expanded and cooled. But on the small scale, gravity pulled bits together. A star is one such gravity-induced concentration of matter.

Think of a star as a place where gravity pulled a small piece of the universe together, where it heated that tiny place up to high temperatures again—high enough to knock electrons off atoms once again. And so parts of the now much expanded universe re-ionizes because of the formation of stars. This added light is detectable today; it very slightly alters the cosmic microwave background in ways that can just barely be untangled. And this allows us to estimate approximately when the first stars formed: the answer is about one half billion years into the 13.8-billion-year expansion of the universe.

These stars did more than re-ionize the matter and make light. They also—at the extreme temperatures of their centers—reignited nuclear fusion. And so the original Big-Bang-originated ratio of 3/4 hydrogen, 1/4 helium was gradually altered. Galaxies also formed, again with gravity pulling things together—taking the uniform smoothness of the early universe and giving it the lumpiness we see today.

6.3.7   FORMATION OF THE MILKY WAY, SUN, AND EARTH

The Milky Way was not one of the earliest of galaxies to form. But like most galaxies, it did form relatively early on in the history of the universe, at roughly 680 million years since the Big Bang—about 13.1 billion years ago.

The Earth formed as part of the process which formed the Sun—although the formation of the Sun was completed slightly before that of Earth. But this happened much after the formation of the Galaxy. The Sun began to form only about 4.60 billion years ago, and the Earth was mostly formed soon after—about 4.54 billion years ago.

By the time the Sun formed, the Milky Way Galaxy had been around a while—about 8.5 billion years. During these eons, multiple generations of stars formed, and used their extreme gravity to heat hydrogen and helium to the enormous temperatures and densities first encountered in the first few minutes of the Big Bang itself. These conditions allow for lighter elements—hydrogen and helium—to undergo nuclear fusion and thus make heavier elements. These stars then added heavier elements to the original *primoridal* mix of 3/4 hydrogen, 1/4 helium produced by the big bang itself. As later generations of stars formed, they were made in part from these heavier elements produced by earlier generations of stars.

Thus, the solar system contains about 2% of "metals"—elements heavier than hydrogen and helium. This is a lucky thing for me! Otherwise, this book would never have been published and you would not be reading it; the complexity of life depends utterly on the elements heavier than hydrogen and helium.

6.3.8   TIME OF DARK ENERGY DOMINATION

"Normal" gravity is attractive, and it is stronger at smaller distances. But there is evidence of another effect that in some ways acts in the opposite sense. It is negligible at small distances, only becoming noticeable at very large, cosmologlcal distances. And its effect is repulsive; it makes the universe expand faster and faster. We call this effect, somewhat whimsically, *dark energy*.

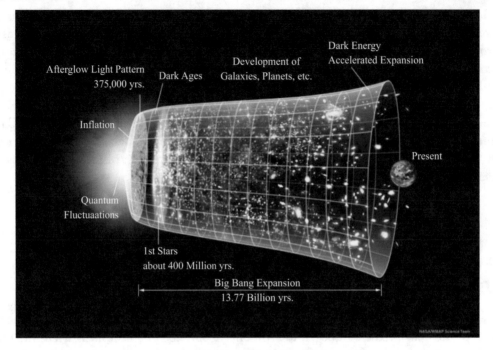

Figure 6.2: A graphical illustration of the history of the universe. The time scale is on neither a linear nor a logarithmic scale. (Graphic credit: NASA/WMAP Science Team.)

Normal gravity was the dominant factor in the early universe, and this made the universe expand at a slower and slower rate. But as normal gravity became weaker, dark energy became stronger. About 4 billion years ago—9.8 billion years from the Big Bang, this dark energy effect became dominant. And so now we see a universe that is expanding *more* rapidly as time goes by.

6.4   A GRAPHICAL SUMMARY

Figure 6.2 shows a graphical depiction of the history of the universe. The illustration shows, through its overall shape, the evolution of the *scale factor*—the changes in the distance between two arbitrarily-chosen co-moving points in the expanding universe.

Notice that the scale factor initially expands at an accelerating rate, and so is concave-outward in the diagram. This is meant to represent the period of cosmic inflation. Afterward, the shape of the diagram is ever-so-gently curved concave inward, representing the decelerating effect of ordinary gravity, slowing the rate of expansion as time passes. But near the end—that is to say, the present—the diagram begins to again expand at an accelerating rate, signifying the onset of the domination of dark energy over ordinary gravity.

This figure, and many others like it, uses a rather whimsical time scale to go horizontally left-to-right from the beginning of the Big Bang to the present. The scale is neither linear nor strictly logarithmic. A linear scale does not work for reasons already discussed; events near the beginning of the big bang, occurring within a fraction of a second, would be indistinguishable set in context with events that occurred billions of years later.

But a logarithmic scale, such as shown in Figure 4.2 depicting the same sequence of events, has its own disadvantages. Such a logarithmic scale hides the fact that much of importance happened right at the beginning, within a fraction of a second. This illustration, on the other hand, attempts to capture the spirit of the timing of the events in the history of the universe, even as it sacrifices numerical accuracy.

6.5   A COSMIC CALENDAR

Instead of putting events in the history of the universe on a long paper tape, we can appeal to our knowledge of the annual calendar. We all know from experience how long a year is; but we also experience directly, and so have an intuitive feel for, tiny fractions of that year—namely months, weeks, days, hours, minutes, and seconds. And so we can, using an ordinary linear scale, put our history of the universe onto a single calendar year, with the beginning of the Big Bang starting the instant after midnight on January 1, and the present just before midnight one year later, on December 31.

Since we use 365 days to represent our 13.8 billion years since the beginning of the Big Bang, then one day represents $13.8/365 = 0.0378$ billion, or 37.8 million, years. Similarly, an hour represents 1.57 million years, a minute 26,300 years, and a second 438 years.

Since we have already calculated a time line to represent the history of the universe in a space of 100 m (Chapter 4, Section 4.1), to put these same events on a 365-day calendar we need simply multiply the numbers in the third column of Table 4.1 by 3.65.

Looking at Table 6.3 emphasizes again that for the history of The Universe, much of the action was at the very beginning, when the universe was only a tiny fraction of a second old. But I have also thrown in some more recent events. The Cosmic Calendar has the important feature of putting ourselves in context; the universe was a done deal, by the time we came along, at only eight minutes before midnight. Even the dinosaurs had to wait until Christmas to have their day.

An excellent review of different uses of historical time lines and variations on the idea of the cosmic calendar can be found at en.wikipedia.org/wiki/Cosmic_Calendar.

Table 6.3: A cosmic calendar compresses the entire history of the universe into one calendar year

Event	t (s)	Date	Time
End of Planck era	1.0×10^{-43}	January 1	0 h, 0 m, 10^{-53} s
End of inflation era	1.0×10^{-35}	January 1	0 h, 0 m, 10^{-43} s
Proton-neutron ratio fixed	1.00	January 1	0 h, 0 m, 10^{-10} s
End of fusion era	180	January 1	0 h, 0 m, 10^{-8} s
Time of decoupling	1.2×10^{13}	January 1	00:14
Re-ionization begins	2.1×10^{16}	January 18	
Formation of Milky Way	2.4×10^{17}	January 21	
Formation of Sun	2.81×10^{17}	August 25	
Formation of Earth	2.89×10^{17}	September 1	
Dark energy era begins	3.00×10^{17}	September 10	
T. rex	4.32×10^{17}	December 27	09:00
First humans	4.35×10^{17}	December 31	23:52
End of last ice age	4.35×10^{17}	December 31	23:59:33
The present (t_0)	4.35×10^{17}	December 31	Midnight

6.6   REFERENCES

Alan H. Guth. Inflationary universe: A possible solution to the horizon and flatness problems. *Physical Review D*, 23:347–356, January 1981. DOI: 10.1103/PhysRevD.23.347 91

CHAPTER 7

The Future

7.1 THE FUTURE AT LARGE SCALES

Figure 7.1 shows the evolution of the universe for several different cosmological models. Unlike Figure 6.1, this is not a Hubble diagram of velocity (or redshift) vs. distance. Rather it is a graph of the relative *scale factor* of the universe plotted vs. time. The scale factor can be thought of as the distance between two arbitrary co-moving observers—located on two galaxies distant from each other, for example. As time passes, this distance increases, and the progression is a record of the expansion of the universe.

The graph extends from the past, through the present ($t = 0$), through projections of the future. The *slope* of the graph at the present, marked by "now," is set by the Hubble constant. This is *observed* to be about 70 km/s per megaparsec, and so it is fixed. Thus, all of the models shown have the same slope at that point. If they did not, they would not agree with our measured value of H_0.

We observe the past, through the concept of look-back time. The future, on the other hand, is a projection. If we apply a particular physical model to the expansion of the universe, does it agree with what we observe of the past? And if so, what does that model predict about the future?

Figure 7.1 shows five possibilities. The straight line marked $\Omega_M = 0$ shows the consequence of a model universe that has neither matter nor dark energy. Projected back to the big bang, it gives an age that agrees well with our observations of the oldest objects in the universe. But it disagrees with observation in other details. In particular, there quite obviously *is* matter in the universe, and an accounting of both ordinary and dark matter indicates there is enough to significantly alter the expansion of the universe.

The model marked $\Omega_M = 1$ has some appeal, because it agrees better with estimates of the total amount of both ordinary and dark matter. It has other agreeable qualities as well, as we shall see later. But the age is wrong; it implies the universe is less than 10 billion years old. And we have good evidence that the globular clusters, as just one example, are significantly older than this.

Four of the five models expand infinitely into the future. But one—the one marked $\Omega_M = 6$—eventually contracts. At about 5 billion years in the future the universe stops expanding and then begins to contract, leading to the *Big Crunch* in about 18 billion years. This is an intriguing idea, but it would require a density of matter many times greater than what is observed, even when dark matter is taken into account.

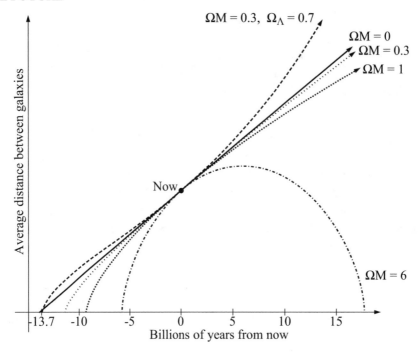

Figure 7.1: Possible futures of the expansion of the universe, as calculated by different cosmological models. (Graphic by BenRG, Public Domain.)

The model that best fits the observed data, and also agrees with other important assumptions that we will explore later, is the strangest of all. It is marked $\Omega_M = 0.3$, $\Omega_\Lambda = 0.7$, and it represents a universe with both dark matter and dark energy. Notice that for all of the models—this one included—the expansion of the universe is initially slowing down. But this dark energy model eventually begins to expand faster and faster, accelerating rather than decelerating.

This more complex model that includes dark energy agrees better with our observations of distant galaxies; the observed look-back time really does seem to have this shape. If this slightly more complex—but still rather simple—model is correct, then it implies the universe will expand forever at an ever-accelerating rate.

7.2   THE FUTURE AT SMALL SCALES

While the universe as a whole is expanding, there are other processes at work on the smaller scales of individual stars and galaxies. The laws of statistics and thermodynamics naturally move matter and energy from a more ordered state to a more randomized arrangement. The molecules in a gas injected into one side of a container randomly collide, quickly diffusing until the entire container is filled, while the opposite process is simply too unlikely to happen. The glass that fell to the floor and shattered does not pull itself together and jump back to the table, even if given enough energy to do so. This basic principle is related to the concept of *entropy*—and it seems to be a basic law that it always increases.

Gravity on the other hand *seems* to work in the opposite sense; it pulls a random cloud of gas and dust into a star, for example. And so gravity appears to impose an order on the chaos. The formation of a star by gravity does not *really* defy the laws of statistics and thermodynamics, but from our perspective it seems to in important ways. Intense light is produced coming from a small place, for example, when a star is formed. And this allows for the possibility of life [Penrose, 2004, Sec. 27.7].

But this is only temporary. As we shall see in Chapter 9, the stars eventually exhaust all possible means for generating energy, and so go dark in the form of some compact gravitational concentration of mass—either a white dwarf, a neutron star or a black hole. It is possible that the universe could eventually be a dark and lonely place—dark, compact points of mass that no longer interact with each other, increasingly separated by an accelerated expanding universe.

But it is important to remember that our best prediction for the future of the universe— the $\Omega_M = 0.3$, $\Omega_\Lambda = 0.7$ curve in Figure 7.1—is the projection of a *model*. It is the simplest model that fits the observations, but other more-complex possibilities are imaginable that would also agree with current observations. And even if the universe does eventually become dark and uninteresting, projections of the $\Omega_M = 0.3$, $\Omega_\Lambda = 0.7$ model suggest it would take an almost unimaginably long time to do so—on the order of 10^{14} years. So it is still important to take out the trash now and then.

Long before the universe itself becomes unrecognizable, enormous changes are certainly in store for Earth. The Sun for example—absolutely necessary for life on Earth—will very much turn against us in roughly 5 billion years. It will become brighter and larger, cooking Earth and eventually swelling so large that Earth will be consumed.

But remember our cosmic calendar. Five billion years is also roughly the age of Earth; and so we could say that Earth is, literally, middle-aged. If we place the history of the universe in a calendar of one year, then 5 billion years occupies roughly the last four months. But the whole record of humanity, by comparison, occupies only the last *8 seconds* of December 31. So it is important to keep a sense of proportion!

7.3   REFERENCES

Roger Penrose. *The Road to Reality: A Complete Guide to the Laws of the Universe*. Vintage Books, 2004. 101

PART III

Evolution

CHAPTER 8

Evolution of the Solar System

8.1 COMPONENTS OF THE SOLAR SYSTEM

The Sun formed about 4.6 billion years ago, and the basic layout of the rest of the solar system formed in that same context. We have already described the sizes and distances to what have historically been called planets, but there are other important types of objects too.

This begs an obvious question. How do we decide whether or not something "belongs to" the solar system? This may seem an arbitrary decision. But there is a specific answer that makes sense: objects in the solar system are *gravitationally bound to the Sun*. This means that solar system objects perenially orbit the Sun, and so cannot escape its gravity.

Even this sensible definition has its limits; objects at the fringes of the solar system are only *weakly* bound to the Sun, and so they can be captured by nearby stars. That caveat aside, I briefly describe in Section 8.1 the most significant parts of the solar system.

8.1.1 TERRESTRIAL PLANETS

The terrestrial (Earthlike) planets are Mercury, Venus, Earth, and Mars. They have many properties in common that are distinct from those of the Jovian planets described in Section 8.1.2. The terrestrial planets are relatively small and have comparatively high average densities, consistent with a composition of rock and metal.

Earth is largest, with Venus not far behind. Mars and Mercury are considerably smaller, with Mercury only 50% larger than Earth's (only) satellite, the Moon. See Figure 8.1 for images made to scale.

Figure 8.1 depicts an Earth telescope view of Mars, but the other three planets are portrayed with imagery from the space program. The picture of Venus is not a picture—it is an elevation map of the surface made with radar from an orbiting space probe; the orange color was chosen arbitrarily, and is not meaningful. Venus has a dense, opaque atmosphere, and only radar can peer through the relatively featureless clouds.

The terrestrial planets have relatively few satellites. Earth has its very-large Moon, and Mars has two tiny satellites; Venus and Mercury have none.

8.1.2 JOVIAN PLANETS

The Jovian planets are Jupiter, Saturn, Uranus, and Neptune. They are very much larger than the terrestrial planets. The smallest, Neptune (Uranus is only slightly larger), has nearly 4 times

Figure 8.1: The terrestrial planets, portrayed to scale. From left to right: Mercury, Venus, Earth, and Mars. (Image credit: NASA/JPL/HST/Mercury Globe-MESSENGER, Public Domain.)

the diameter, roughly 60 times the volume, and over 17 times the mass of Earth. Jupiter, the largest, is 11 times Earth's diameter, has over 1400 times its volume, and over 300 times Earth's mass. But the average *densities* of the Jovian planets are much lower than those of the terrestrial planets. Jupiter, for example, has a density less than 1/4 that of Earth. Saturn is less dense, on average, even than water.

This means the Jovian planets must be made primarily of the lightest of elements. The largest, Jupiter and Saturn, are made primarily of the two lightest elements, hydrogen and helium. See Figure 8.2 for a to-scale depiction of the interiors of the Jovian planets. They are compressed by gravity to a liquid, except for the outermost atmospheric layer of gases. There is likely a (relatively) small core of rock and ice. Uranus and Neptune are slightly denser, with greater proportions of rock and ice compared to hydrogen and helium.

8.1.3   ASTEROIDS

The asteroids are small, mostly rocky bodies that have many different orbits, but the majority lie between Mars and Jupiter in what is known as the *asteroid belt*. See Figure 8.3 for a graphical depiction of the orbits of many thousands of known asteroids. The bright white marked orbit is that of Earth, and it is clear that most lie in the asteroid belt between Mars and Jupiter.

The caption of Figure 8.3 contains a link to an animation of these orbits, showing that they do progress all in the same sense, and this is the sense of orbits in the solar system in general. But close inspection shows that the asteroids form a thick torus of orbits, and those of many are inclined at least several degrees to the plane of the rest of the solar system.

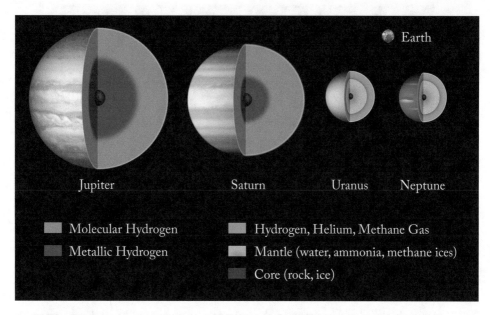

Figure 8.2: The Jovian planets shown to scale, with their interior structures. Earth is shown to scale for comparison. (Graphic credit: Lunar and Planetary Institute, Public Domain.)

Figure 8.3: The plotted orbits of many thousands of asteroids. The white orbit is that of Earth; most of the asteroids are in the belt between the orbits of Mars and Jupiter. Possible Earth-crossing asteroids are shown in blue. Video of asteroid orbits, with near-Earth asteroids highlighted. (Image credit: NASA/JPL-Caltech.)

Figure 8.4: The asteroid Mathilde. It is about 50 km across. (NASA, Public Domain.)

The asteroids highlighted in green in the animation are dubbed *near-Earth asteroids*. Their orbits potentially could cross that of Earth's, and so there is at least a slight risk of collision (albeit far in the future).

The largest asteroid is Ceres, at about 950 km in diameter. The second-largest is Vesta, only a bit more than half that size. All of the rest are too small for gravity to pull them into spheres, and so they are essentially random shapes. Some asteroids have been visited with space probes. See Figure 8.4 for a close-up image of one.

8.1.4   OORT CLOUD COMETS

Most comets are small (only several kilometers in diameter), mostly icy bodies that spend the bulk of their time too far from the Sun to be seen. The largest reservoir of these bodies is far beyond the orbits of the planets, nearly one fourth of the distance to the nearest star. This mostly spherical shell of cometary objects is called the *Oort cloud*.

The Oort cloud cannot be seen directly; its existence is inferred by modeling the interaction of icy bodies with the newly-formed Jovian planets in the early solar system, and by studying the orbital dynamics of the many individual comets we *do* see. For comets are only visible if they come relatively close to the Sun, which then vaporizes some of the ices on their surfaces, forming a large and visible glowing cloud around the tiny nucleus. This process sometimes also releases dust, which reflects sunlight.

On the occasion that a comet does come to the inner solar system, the large cloud of gas and dust released by the Sun's radiation is sometimes easily visible to the naked eye. See Figure 8.5 for an example—comet Hale–Bopp that graced the skies in 1997.

Figure 8.5: Comet Hale-Bopp as photographed with a small telescope. The stars in the background appear as short streaks because the telescope was tracked instead on the tiny nucleus of the comet. It's orbital motion about the Sun is apparent even over the course of this 30-min exposure. Photograph by the author.

Oort cloud comets have plunging orbits that can come from any direction; their orbits are often not at all aligned with the plane of the solar system. They quickly swing around the Sun and are never seen again. These comets, with their highly elongated orbits, spend the overwhelming majority of their time moving slowly, very far from the Sun.

8.1.5   PERIODIC COMETS

A small minority of comets have orbits that are relatively small—with periods from several years to several hundred years. And so we see such a comet repeatedly throughout human history. Most of these *short-period comets* have very elongated orbits, that take them to the outer planets or even beyond.

Comet Halley is a good example, with its orbital period of roughly 76 years. For the vast majority of that time, it is too far away for the Sun to evaporate a visible cloud of gas and dust around it. But its highly elliptical orbit brings it periodically close to Earth and the Sun, and we get a spectacular view. See Figure 8.6 for one of the few close-up images of the nucleus of a comet, taken from the Deep Impact space probe.

Regarding periodic comets, however, Halley is the exception. Most are far from spectacular to the naked eye, with most only visible in telescopes, even when at their brightest.

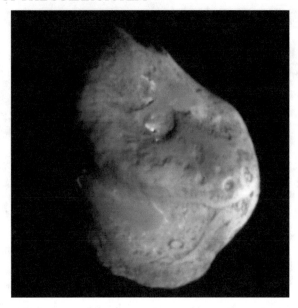

Figure 8.6: The nucleus of comet Tempel 1. (Image credit: NASA/JPL-Caltech/UMD, Public Domain.)

8.1.6 METEORS AND METEORITES

Small, solid bits of solar system stuff often collide with Earth's atmosphere. The intense heat generated by friction with Earth's upper atmosphere causes the air along the particle's trajectory to glow with a visible trail called a *meteor*. Since these particles impact the atmosphere at relative speeds of tens of kilometers per *second*, and the substantial kinetic energy that comes with that, even a pea-sized particle can make a bright meteor.

Most meteors have their origin in the short-period comets, their nuclei broken up by close passages with the Sun. These relatively insubstantial pieces typically vaporize (or are reduced to dust) high in the atmosphere.

But larger pieces of rock and metal—mostly broken pieces of asteroids that have collided with each other—can make it all the way to the ground. These *meteorites* can then be studied in the laboratory; studies of meteorites provide much of the foundation for our knowledge of the asteroids. See Figure 8.7.

8.1.7 KUIPER BELT OBJECTS (KBOS)

The Kuiper belt is a distant region of small solar system bodies that are mostly a mix of rock and ice. The Kuiper belt is much closer than the Oort cloud; it begins at about the orbit of Neptune. Like the asteroid belt, the Kuiper belt is aligned somewhat with the plane of the orbits of the planets.

Figure 8.7: Left: an acid-etched slice of an iron meteorite. Right: a cut and polished piece of a pallasite meteroite. Photographs by the author.

Pluto is widely believed to be simply a relatively nearby and large Kuiper Belt Object (KBO). Many of the short-period comets are also believed to have their origins there.

8.1.8   SATELLITES

There are many *satellites* or *moons* orbiting the planets, nearly all accounted for by the moons of the four Jovian planets. For an excellent graphic illustration, too large to print in this book, see https://commons.wikimedia.org/w/index.php?curid=42598104.

Self-Gravitating Satellites

There are a number of large satellites of the planets, massive enough that gravity pulls them into spheres, and *differentiates* them into a more-dense core surrounded by a less-dense mantle. Several of these satellites (including our own Moon) are larger than Pluto, and one (Ganymede, which orbits Jupiter) is larger than Mercury. The systems of the larger satellites around the Jovian planets mimic the solar system as a whole. The moons orbit in nearly the same plane, aligned (and in the same sense) as the rotation of the planet.

Small Satellites

The Jovian planets have dozens of satellites too tiny for gravity to pull them into spheres. Some of these have orbits that do not match up with the orbits of the larger satellites. All three moons of the terrestrial planets are rocky. The satellites of the Jovian planets are mixtures of rock and ice, with Jupiter having more rocky satellites and Neptune having more icy satellites.

8.2   OVERALL PROPERTIES OF THE SOLAR SYSTEM

A theory of the formation and evolution of the solar system should account for the many different types of objects that orbit the Sun. But there are many other basic facts about the solar system that must also be acknowledged—and explained.

1. With only a few exceptions, the orbits of the planets, their rotations, the orbits of their satellites, and the orbits of the major asteroids all lie within roughly the same plane, and this plane roughly matches the equator of the Sun (and so aligns with the Sun's rotation).

2. The orbits and rotations of most solar system bodies go round in the same sense—counterclockwise as seen from above Earth's north pole—and this sense is the same as the rotation of the Sun.

3. There is a trend in composition from the inner solar system to its outer reaches. Heavy elements, such as metals like iron and aluminum, and the building blocks of rocks, such as silicon, are more common (as a percentage) closer to the Sun. Lighter elements, such as hydrogen, helium, carbon, nitrogen, and oxygen, form a greater percentage far from the Sun.

4. The atmospheres of the terrestrial planets (excluding Mercury, which has no significant atmosphere) are made mostly from molecules that do *not* contain hydrogen; water is a prominent exception to this rule. The atmospheres of the Jovian planets, on the other hand, are—again, apart from water—made of molecules that do not contain *oxygen*.

5. The orbits of the terrestrial planets are relatively close together, while those of the Jovian planets are much further apart from each other.

6. The asteroid belt is located in the boundary region between the terrestrial and the quite-different Jovian planets. Is this merely a coincidence?

7. The orbits of the larger bodies in the solar system are roughly circular. Smaller bodies often have orbits that are noticeably elliptical, but highly elliptical orbits are uncommon, and tend to belong to objects with very long periods of orbit about the Sun.

8. Some of the solid objects in the solar system are covered with impact craters; the Moon is a good example. Other objects—Earth for example—have relatively few identifiable impact craters. There is a strong correlation with size; larger solid bodies tend to have fewer visible impact craters than smaller solid bodies.

9. The Jovian planets all have many satellites each, and their systems of satellites mimic in some ways the solar system as a whole. This is in marked contrast to the terrestrial planets. Neither Mercury nor Venus has any satellites at all, Earth has only one—the Moon—and Mars has only two, both of which are tiny.

10. There are individual exceptions to these basic patterns observed for the properties of the solar system. Are there at least plausible scenarios that explain these exceptions?

8.3   THE NEBULAR HYPOTHESIS

The basic outline of a plausible scenario for the formation of the solar system has been known for centuries. The basic concept behind the modern *solar nebular model* was first put forth by Immanuel Kant in 1755, and in more detail about 40 years later by the French mathematician Pierre–Simon Laplace.

A large cloud of gas and dust was pulled together by its own self-gravity, becoming hotter and denser toward its center as it contracted. Most of the contracting mass became the Sun; we will explore that process in more detail in Section 9.3.

Even a slight initial rotation of this contracting cloud would have been greatly amplified because of the *conservation of angular momentum*. One of the fundamental principals of physics, it is why an ice skater spins faster when they pull their arms inward, closer to their rotation axis. But for a cloud of gas and dust contracting due to its own self-gravity, there is another consequence—some of the cloud would flatten out into a thin disk of material orbiting the newly-formed star.

The planets then formed out of this *protoplanetary disk*. The nebular hypothesis explains in a natural way the fact that the planets orbit in the same direction, in very nearly the same plane. Furthermore, there is evidence of such protoplanetary disks around at least some newly-forming stars. See Figure 8.8 for an example—the star HL Tauri as imaged by the Atacama Large Millimeter Array (ALMA), in Chile. The gaps in the disk are just what one expect if there are proto-planets—too small to see—clearing out spaces in the disk as they orbit.

But this all begs the question of just how do we end up with planets? What causes them to form out of the gas and dust? We take up this issue in Section 8.4.

8.4   THE CONDENSATION SEQUENCE

The inevitable consequence of a cloud contracting because of its own gravity is that the densest parts, closer to the center, are hotter. As the proto-planetary disk cooled, there was a gradation of temperature, with the hottest parts being near the newly-formed Sun, the coolest parts in the outer regions of the disk.

As the temperature decreased, tiny solid particles and liquid drops could condense out of the solar nebula. Little pieces could then stick together to form bigger pieces. The bigger a clump is, the more efficiently it grows by sweeping up new material. And so larger clumps grow at the expense of smaller clumps. If a clump becomes massive enough, then its own self gravity will hold it together and gravitational attraction greatly accelerates this *accretion* process.

But the chemical composition of these condensing solid or liquid particles depends critically upon the temperature and pressure. The details are complex, but the basic trend is fairly simple:

Figure 8.8: The protoplanetary disk around the star HL Tauri, as imaged by the Atacama Large Millimeter Array, in Chile. (Image credit: ALMA ESO/NAOJ/NRAO, CC BY 4.0.)

Molecules made of heavier elements tend to condense out at higher temperatures, and so condensed out of the solar nebula first. Molecules made of lighter elements condensed out later, and further from the Sun where the solar nebula was cooler.

There is another issue: as molecules were condensing out of the solar nebula, what was there to work with? What were the most common elements capable of making solids or liquids? We already have this answer—it is the roughly 2% of "metals" that makes up the Sun and interstellar clouds of gas and dust today. These elements can potentially combine with each other and with hydrogen to form molecules that can take solid or liquid form. Helium does not enter into the calculation, as it does not chemically combine with other elements.

And we have already seen that the "metals" are made of mostly lighter elements; *carbon, nitrogen, and oxygen are the most prominent* (see Section 5.4.8). After that, the elements silicon, aluminum, magnesium, calcium, iron, nickel, and sulfur are important.

There are four basic categories of materials that can be used to form planets. I list these basic building blocks as follows, in order of decreasing abundance and increasing density:

1. Gases: the bulk of the solar nebula was hydrogen and helium. Under the right conditions, the hydrogen will combine with itself to make H_2 molecules. Helium combines

with nothing. These two lightest elements are gaseous unless compressed to a liquid by high pressures. But hydrogen and helium can be used to make planets only if there is enough gravity to hold them together.

2. Ices: carbon, nitrogen, and oxygen will combine with hydrogen to make molecules that can be either gaseous or solid. Water, H_2O, is the most abundant ice in the solar system, and it is what both hydrogen and oxygen will form if given the chance. If there is not enough hydrogen, the remaining oxygen can then form molecules such as CO_2 (carbon dioxide). If on the other hand there is not enough oxygen, the remaining hydrogen can form ices such as methane (CH_4) or ammonia (NH_3).

3. Rock: rocky materials are primarily made of combinations of oxygen, silicon, aluminum, iron, magnesium, and calcium. Olivine, a magnesium iron silicate, is a good example; it is the primary component of Earth's rocky mantle, and is also found in meteorites. The greenish-yellow crystals in the meteorite on the right in Figure 8.7 are made of olivine.

4. Metal: that is to say, real metals such as iron and nickel that can form metallic alloys, sometimes in combination with sulfur.

The condensation sequence means that *ices could condense out of the solar nebular only at temperatures much lower than for rocky materials, which condense out at temperatures less than metals such as iron and nickel.* Thus, very close to the Sun, only rock and metal condensed out of the solar nebula.

Far from the Sun, ices could condense, especially the most abundant—ordinary water ice. Rock and metal condensed out too, just as it did close to the Sun. But the fact that water could condense out in the outer solar system means that massive *proto-planets*—probably several times the mass of Earth—quickly developed out of this mostly icy material. These icy pre-planetary bodies would have likely had cores of rock and metal, but the bulk of the mass was ice.

And so we have a scenario in which the Jovian planets began with these large proto-planetary cores that had enough mass to pull in their own self-gravitating piece of the solar nebula—creating what looks like mini-versions of the solar system as a whole. Thus, each Jovian planet formed from its own self gravitating concentration of mostly hydrogen and helium at the center of an orbiting disk of gas and dust. The major satellites of Jupiter, Saturn, Uranus, and Neptune then formed from the disks around these newly formed planets.

This process of a cloud of gas pulling together under its own self gravity is strongly dependent on temperature; if the temperature is too high, the cloud will be instead supported by gas pressure and will not collapse. And so the inner planets, with smaller cores forming in a hot part of the solar nebula from only much-less-abundant rock and metal, could not use gravity to pull in their own massive envelopes of hydrogen and helium.

8.5   THE LATE HEAVY BOMBARDMENT

There are many individual exceptions to the overall properties of the solar system, as outlined in Section 8.2. Uranus rotates, and its moons orbit, at an angle nearly perpendicular to the plane of the solar system. Venus rotates very slowly, and backward. Earth has a large Moon, unlike the other terrestrial planets.

How might these exceptions have come about? A glance at the surface of our own Moon provides a clue. It is covered by *impact features*; nearly every visible mark on the Moon is at least indirectly due to high-speed collisions with smaller bodies. With only a few exceptions, all of the smaller solid bodies (and most of the larger ones) in the solar system bear the scars of impact features.

The most obvious impact features are craters—nearly circular marks that are either bowl-shaped (if they are small), or consist of a circular rim of material surrounding a shallow depression (sometimes with a central mountain peak). This is just the feature one expects from an impact with an object moving at high velocity. The kinetic energy of the impacting body creates an enormous explosion, and it is this explosion that makes the crater. The crater thus formed is far larger than the impacting body, which is destroyed in the impact.

If the impacting body is large and energetic enough, a huge *impact basin* may be formed. The large darker areas on the Moon that mark its visible face-like features are the remains of such impact basins. These *lunar maria* regions are darker, flatter and show fewer craters than the *lunar highlands* regions that surround them. The highlands, on the other hand, show evidence of craters on top of craters, covering still more craters.

And so we can hypothesize a period of intense bombardment that formed the highlands regions of the Moon's surface. Near the end of this period, in the *late heavy bombardment*, the Moon suffered several giant impacts that created the lunar maria, destroying the earlier impact craters in those regions. After that, the rate of impacts decreased dramatically, leaving the younger (but still very old) lunar maria regions relatively free of craters.

The Apollo missions to the Moon brought back many rocks from mostly the lunar maria, but also the highlands—and these rocks date from the early solar system, roughly 4.5 billion years ago. The most plausible scenario for the formation of the Moon is the *giant impact theory*: the Moon formed as a result of a giant impact between proto-Earth and a (no longer existing) Mars-sized proto-planet.

Thus, we have an early solar system plagued by collisions between the newly formed bodies. This would have caused a period of intense bombardment of the surviving solar system objects—both planets and moons. But eventually, most everything that could collide, would have already collided, making collisions increasingly rare. Models of this process suggest the period of intense bombardment would have mostly ended with a few very large impacts, in rough agreement with the observations of the lunar maria.

Today, impacts occur at a far smaller rate—fortunately! Since the vast majority of the impacts happened billions of years ago, early in the solar system's formation, only bodies with an ancient surface bear the obvious scars of this period of intense bombardment. The surface features of Earth, for example, are dominated by the remnants of *internal processes*; the crust of Earth is constantly altered by plate tectonics (continental drift) and volcanoes. And so most of the original impact features have been long covered over; only the hundred or so most recent ones are left. Smaller bodies, such as the Moon, did not have enough internal heat to greatly alter their surfaces with plate tectonics, because they cooled off faster. And so the Moon still bears the scars of nearly its entire history.

8.6   FORMATION OF THE COMETS AND ASTEROIDS

Small bodies in orbit about the Sun are highly susceptible to having their orbits greatly altered (*perturbed*) by the gravity of the large planets. Jupiter, by far the most massive of the planets, is the most common cause of these perturbations. In the region between Mars and Jupiter, no large rocky planet could form; the orbits there are constantly perturbed by Jupiter's gravity such that they cross over each other. These crossed orbits in the asteroid belt region result in violent collisions that break bodies apart more often than allowing them to join together to make larger bodies.

Beyond the asteroid belt, icy bodies formed in the early solar system. It is hypothesized that repeated encounters with the enormous gravity of Jupiter deflected most of these small icy bodies out of the part of the solar system occupied by the Jovian planets. Some of them would have been deflected to the inner solar system, adding icy bodies to the mix of impacts suffered by Earth, the Moon and the other terrestrial planets. This is considered a likely origin for the water on Earth; our planet formed too close to the Sun for water to have condensed out directly.

But models suggest that many of the small icy proto-bodies that condensed out of the solar nebula in the region of the Jovian planets would have been deflected to much larger distances, to form the Kuiper belt and the Oort cloud (see Sections 8.1.7 and 8.1.4).

CHAPTER 9

Stellar Evolution

9.1 M, L, R, AND T

When we look out into space, we see only a snapshot of the life histories of stars. Fortunately for astronomers, there are *many* stars, representing nearly all stages in stellar evolution. There are four basic overall properties of stars that we can observe and measure.

- **Mass (M):** We measure the *masses* of stars in *binary star systems*, by seeing the effect gravity has on their orbits. Instead of using kilograms, we describe the masses of stars as they compare to the Sun, which we represent with the astrological symbol $\odot$. A star with 10 times the mass of the Sun, for example, has a mass 10 $M_\odot$.

- **Luminosity (L):** *Luminosity* is the total amount of energy emitted per second by a star. We determine this by measuring both the apparent brightness, as seen from Earth, combined with the distance to the star. As is the case for mass, we use $L_\odot$ to represent the luminosity of the Sun.

- **Radius (R):** The *radius* can be measured *directly* for only a few stars, as most appear as only points in even the most powerful of telescopes (see Section 9.1.1). We use the symbol $R_\odot$ to represent the radius of the Sun.

- **Temperature (T):** The spectrum of a star includes much information about the *temperature* of the gases that are emitting its light. Numerically, temperatures of stars do not vary as much as luminosities, or radii, and the numbers as measured on the absolute Kelvin scale (K), are typically only in the thousands or tens of thousands. And so we simply denote the temperature of a star on this physical scale. As an example, the Sun has a temperature of about 5800 K. When we say this, we mean the temperature only in the thin layer of gases where the light gets out—it describes the physical nature of the light we *see*. The opaque inside of the star is a different matter altogether.

9.1.1 THE RELATION BETWEEN L, T, AND R

A star emits light because it is hot. More specifically, a star emits most of its light from a narrow range of layers of gas, where the light finally escapes. And it is the temperature of this gas in what is known as the star's *photosphere* that is responsible for much of the details of the emitted light.

We will discuss the details further in Section 12.2.3, but the following simple relations are *approximately* true.

1. All else being equal, hotter stars have a higher luminosity.

2. All else being equal, larger stars have a higher luminosity.

3. Regarding the luminosity of a star, both size and temperature matter, but temperature matters more. Double its radius and a star is four times brighter. But double its temperature and it will be *sixteen* times brighter.

A good example is the double star Albireo, in Cygnus the Swan. Through even a small telescope one sees a beautiful pair of stars—a brighter yellow one and a noticeably dimmer blue one. We shall see (in Section 12.2.4) that the difference in color means the blue star is significantly *hotter* than the yellow one. And so it must be the case the the that *the yellow star is larger*. If they were the same size, the blue (hotter) star would be brighter. And so the cooler, brighter star must be larger to make up for its lower temperature.

The approximate relationship between L, T, and R for stars can be easily captured mathematically. The power emitted *per surface area* is proportional to the fourth power of the temperature. And the surface area is proportional to the square of the radius (the surface area of a sphere is $4\pi R^2$). Thus, the luminosity, $L_\star$, of a given star is simply the product of the area and the power per area:

$$\frac{L_\star}{L_\odot} = \left(\frac{R_\star}{R_\odot}\right)^2 \left(\frac{T_\star}{5770}\right)^4. \tag{9.1}$$

I have expressed Equation (9.1) so as to compare a particular star's L, T, and R to that of the Sun, which has an *effective temperature* of 5770 K. This is the temperature the Sun would have if, given its radius and luminosity, it emitted its light with a perfect thermal blackbody spectrum (see Section 12.2.3). The Sun and other stars emit light only approximately as a blackbody, and so Equation (9.1) can be taken as only approximate for real stars. Alternatively, it can be used as the *definition* of the effective temperature of a star.

9.2   THE HERTZSPRUNG–RUSSEL DIAGRAM

One of the most important tools for understanding the relationships between different stars—and the changes that an individual star undergoes as time passes—is called the *Hertzsprung–Russel diagram*, or simply *H-R diagram*. In its traditional form, it is a graph with luminosity on the vertical axis and *spectral type* on the horizontal axis. Stars are then placed on the graph accordingly, each point representing the measured values for an individual star.

We discuss spectral type in detail in Sections 12.2.7 and 13.2.3; it relates quite directly to *temperature*. But there is a complication, and it comes from a fluke of history; the sequence of spectral types was discovered before its physical cause was understood. And this sequence in its traditional order represents, as it happens, *decreasing* temperature. For this reason, the horizontal scale of the H-R diagram puts the hottest stars on the left side of the diagram, and the coolest stars on the right.

Part of the power of the H-R diagram is that although only temperature and luminosity are directly plotted on the diagram, a third property is at least approximately implied—the radius of the star. As described in Section 9.1.1, there is a relation between luminosity, temperature and radius. This means that if any two of the three are known, the third can be calculated. And so we have the following for the H-R diagram.

1. Stars on the left of the diagram are hottest; those on the right are coolest.

2. Stars at the top of the diagram have the highest luminosity; those at the bottom have the lowest.

3. The largest stars are in the upper-right corner. They are simultaneously high luminosity, but low temperature. The only way for both to be true is if they are very large. Similarly, the stars on the lower-left are the smallest.

Figure 9.1 shows the H-R diagram for many well-known stars. Notice the logarithmic scale of luminosity on the vertical axis, and the scale of decreasing temperature on the horizontal axis. The luminosities are given as multiples of $L_\odot$, and so "1" marks the luminosity of the Sun. The horizontal axis is also labeled with the traditional spectral types, O, B, A, F, G, K, and M (the origins of these letters will be discussed in Section 12.2.7).

When L and T are measured for many stars and plotted on the H-R diagram, the pattern is clearly *not* random. Rather, there are regions of the diagram with many stars, while other areas have no stars at all. This means that some combinations of temperature and luminosity are very common, while other combinations are very rare. The basic regions are marked on the diagram with their common names. The origin of some of the names should be clear; the largest stars are in the upper right, while the smallest are to the lower left. So it should be unsurprising that supergiants and white dwarfs are, respectively, at these locations.

The overwhelming majority of stars, when selected from a random sample of a given volume of space, are located on the *main sequence*, which curves from upper left to lower right. Stars at the upper left are both hot and bright, while those at the bottom left are both cool and dim. This is just as one would expect from temperature alone. And so it is not clear simply by glancing at the H-R diagram whether the stars on the *upper main sequence* are bigger or smaller than those on the *lower main sequence*. But when one looks at the actual numbers, upper main-sequence stars are larger than lower main-sequence stars. The diagonal lines in Figure 9.1 show combinations of temperature and luminosity that result in a constant radius, according to Equation (9.1).

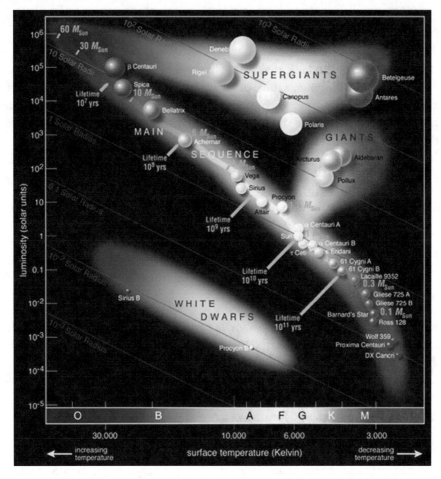

Figure 9.1: The H-R diagram for many prominent stars. (Graphic by ESO, CC BY 4.0.)

9.3   FORMATION OF STARS

We have already seen that stars tend to form in clusters, from gravity squeezing together an interstellar cloud of gas and dust. The Orion Nebula (M 42 or NGC 1976) is an excellent example of such a stellar nursery. See Figure 9.2.

Figure 9.2 combines imagery from both visible and infrared light, to better show its mixture of glowing gas and absorbing dust. Most of the visible light is coming from the outer edge of the glowing cloud; the dust prevents one from seeing through it. The infrared light, however, penetrates the dust somewhat. Spectral analysis of the glowing gas is consistent with a composition of the same basic mix of chemical elements as is the Sun—roughly 75% hydrogen, 25% helium, and about 2% "metals."

Figure 9.2: The Great Nebula in Orion, M 42 (NGC 1976). This is a huge and detailed mosaic made from observations by the Hubble Space Telescope. (Image by NASA, ESA, M. Robberto (Space Telescope Science Institute/ESA) and the Hubble Space Telescope Orion Treasury Project Team, Public Domain.)

Near the center is a tight group of four stars called *Trapezium* (see Figure 9.3). These upper main-sequence stars are all within a couple of light years of each other, and the ultraviolet light they emit provides much of the energy for the glowing gas of the Orion nebula. At infrared wavelengths, many more stars can be seen in the vicinity. The Trapezium stars are the brightest members of an open star cluster in its infancy.

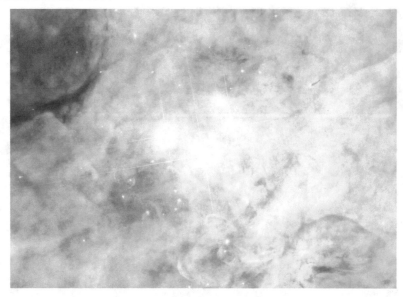

Figure 9.3: The core of the Trapezium star cluster in the Orion Nebula, showing the four highly-luminous and hot Trapezium stars. (Image by NASA, ESA, M. Robberto (Space Telescope Science Institute/ESA) and the Hubble Space Telescope Orion Treasury Project Team, Public Domain.)

Figure 9.4 shows a small detail near the Trapezium. I have enhanced the contrast somewhat, and one can see several small objects that look something like comet-like tails emanating from small bright clumps. These are *evaporating gaseous globules*, or *EGGs*. Presumably, these are stars in the very act of forming. A small part of the nebula has collapsed under its own gravity, and is contracting, making a small, dense clump of gas. One cannot see inside of the EGG because the dust is too thick even for infrared light.

Notice that all of the "tails" point *away* from the Trapezium. Light can create a small pressure on dust, and the Trapezium stars are emitting a lot of light—each is far more luminous than the Sun. This pressure blows dust away from the stars, and the dust carries the gas with it (because of electrical forces). And so the very formation of highly luminous stars in a nebula eventually blows the gas and dust away from those stars—leaving the visible star cluster behind. This also means that stars form only for a limited amount of time. These highly luminous newly formed stars blow away the very gas and dust from which they formed, and so prevent new stars from forming.

In pictures taken with light of even longer wavelengths, further into the infrared so as to peer more deeply into the dust, it is clear that there are hundreds of low-luminosity stars surrounding the four upper-main-sequence Trapezium stars. Apparently, that small part of a much-larger cloud of gas and dust—the Orion Giant Molecular Cloud—began to collapse under

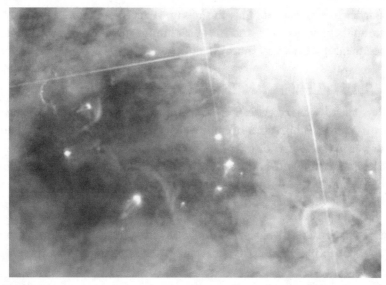

Figure 9.4: An enlarged detail of the Orion Nebula, near the Trapezium. EGGs, tiny comet-shaped structures, can be seen. Small dense clumps containing proto-stars are being evaporated by the intense light pressure of the Trapezium stars. (Image by NASA, ESA, M. Robberto (Space Telescope Science Institute/ESA) and the Hubble Space Telescope Orion Treasury Project Team, Public Domain.)

its own gravity. As it contracted, it fragmented into smaller contracting parts, some of which formed individual stars. The brightest of these stars partly blew a hole in the near edge of the nebula, allowing us to see inside. The ultraviolet light from these very hot stars excited the gas in the nebula, causing it to glow.

9.4   THE MAIN SEQUENCE

There is a physical reason that most stars, when their temperatures and luminosities are plotted on the H-R diagram, tend to lie along the main sequence. For any given star, *most of its visible lifetime* is spent with a luminosity and temperature that places it on the main sequence of the H-R diagram. Since stars spend roughly 90% of their time in this stage, it follows that just by chance, we are most likely to find them as main sequence stars.

A given star changes very little while it is on the main sequence. Most main sequence stars increase their luminosities only very gradually and very slightly, while also becoming slightly cooler. This causes them to gradually migrate roughly perpendicular to the main sequence band (the exact path varies somewhat for different parts of the main sequence). And so the main sequence is a band rather than a narrow line. Ten billion years are required for a star like the Sun to undergoes those rather minor exterior changes.

Thus, a particular star stays roughly in the same part of the main sequence for most of its life. After that, rapid changes ensue, and the star's luminosity and temperature no longer place it on the main sequence; we take up those changes later.

9.4.1   MASS AND THE MAIN SEQUENCE

If individual stars spend most of their lives near only one part of the main sequence, why is it a long *band*, with many combinations of luminosity and temperature, rather than only a small region? The answer is that when different individual stars form, they begin their lives with a combination of temperature and luminosity that places them at *different* locations along the main sequence. The key physical attribute that decides where on the main sequence a star will spend 90% of its visible life is *mass*:

> Stars that form with a relatively large initial mass end up on the upper main sequence, while those that form with a relatively small initial mass end up on the lower main sequence.

And so the main sequence is a progression not only of luminosity and temperature, it is also a sequence of mass. The most luminous and hottest main sequence stars have masses of roughly $50\,M_\odot$, while the least luminous and coolest have masses of only about $0.1\,M_\odot$.

9.4.2   MAIN SEQUENCE LIFETIME

Whatever is the specific "fuel" for the source of a star's luminosity, one would expect that a star of greater mass has more of that fuel. From this isolated observation, one might expect that stars of greater mass have a longer *main sequence lifetime*; it should take longer for them to run out of fuel simply because they have more of it.

But luminosity is equally important, and it turns out to have a far larger effect. The luminosity is the *rate* at which the star emits energy. And so a star with a higher luminosity needs proportionally *more* fuel to emit light for the same amount of time. Put another way, for a given amount of fuel to burn, higher luminosity stars burn it faster, and so should have *shorter* lifetimes.

But we have seen that the higher-mass main sequence stars also have higher luminosities. So which is the bigger effect? The answer is clear if we note that a star 50 times the mass of the Sun (50 times as much fuel to burn) has a luminosity *one million times* greater (it burns that fuel 1 million times faster). And so the high-luminosity stars on the upper main sequence have much *shorter* main sequence lifetimes.

To summarize:

> A star high on the main sequence is hotter, *much* more luminous, more massive, larger, and has a *much* shorter lifetimes than a star low on the main sequence.

9.5   EVOLUTIONARY TRACK OF THE SUN

When a star runs out of its main power source, it changes rapidly. We can follow these changes, plotting both temperature and luminosity on the H-R diagram as time passes, to make a curving *evolutionary track.*

Of course, we can't use a telescope to literally gather such data for an individual star; for most stars, we would have to watch it for billions of years. What we observe for a particular star is simply a point on the H-R diagram, representing what the star is like today (or in the relatively recent past, depending on the look-back time of the star). But if we understand the physics of why stars are the way they are, we can use that understanding to calculate what such a *theoretical* evolutionary track of the star would look like.

Figure 9.5 shows the theoretical evolutionary track for a star like the Sun, which lies roughly in the middle of the main sequence. What do I mean by "like the Sun?" It means a theoretical calculation for a star that had the same chemical composition and, most importantly, the same *mass* as the Sun when it first became a main sequence star. We call this the *initial mass* of the star.

Mass is the key factor that decides the particular changes a star goes through; composition is an important but much-smaller secondary factor. For most stars, the mass stays roughly constant throughout its visible lifetime. We shall see that this is not at all true for the final stages of the life of most stars, and so evolutionary tracks often do not include those final stages. But with these caveats aside, the theoretical track of a star is mostly determined by its initial mass.

Figure 9.6 shows a more schematic evolutionary track of the Sun, and the black line labeled "zero age main sequence" is more affectionately known as the *ZAMS* (pronounced "zams"). It represents the theoretical locations on the H-R digram for stars of *all* masses, just as they arrive at the main sequence, and it occupies the left-lower edge of the observed broad band of the main sequence. This represents the temperature and luminosity of a star, given its mass, when it first becomes stable, fusing hydrogen to helium in its core.

We shall explore in more detail why stars shine in Chapter 15. But the simple answer is that *main sequence stars fuse hydrogen into helium in their cores*; this is the source of energy that allows them to support themselves against their own self gravity, and also to emit an enormous amount of light over millions or even billions of years.

Upper-main-sequence stars, with their enormous luminosities, fuse that hydrogen rapidly, and so they run out of it relatively quickly. For the Sun, in the middle of the main sequence, this *core hydrogen burning* phase lasts about 9 billion years. During that time it changes very little on the outside, moving only slightly to the upper right, almost perpendicular to the ZAMS. But the highest luminosity main sequence stars run out of this hydrogen in their cores in only a million years.

After the hydrogen runs out in the core, the star changes rapidly, swelling to enormous size to become a red giant, and moving far to the upper right on the H-R diagram. A star like the Sun spends about a billion years in this phase, only one tenth its main-sequence lifetime. So

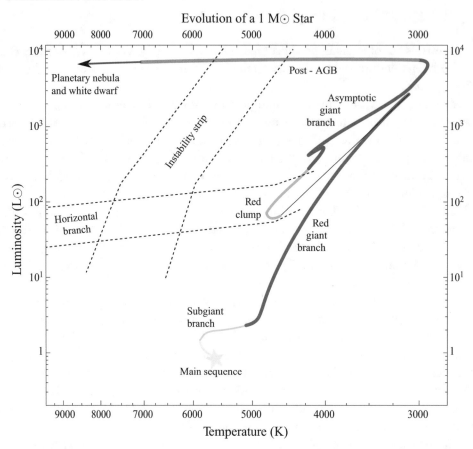

Figure 9.5: Calculated evolutionary track of the Sun, from the moment it begins to undergo hydrogen fusion to the end of its fusion-powered visible life. (Graphic by Lithopsian—Own work, CC BY-SA 4.0.)

catching a solar-type star in the act of being a red giant is less likely than finding it as a main sequence star.

The rest of the changes are far more rapid still, as shown on the diagram. Very briefly, a star like the Sun will become nearly as large as a red supergiant like Betelgeuse, in a phase called the asymptotic giant branch (AGB). Such AGB stars are rare, because any individual star spends very little time (relatively) in this phase.

After the AGB phase, a star like the Sun will become unstable and gently eject its outer layers, losing much of its mass. The exposed core will eventually contract to a white dwarf, about the size of Earth. The white dwarf will then gradually cool off, becoming dimmer until it is no longer visible in telescopes.

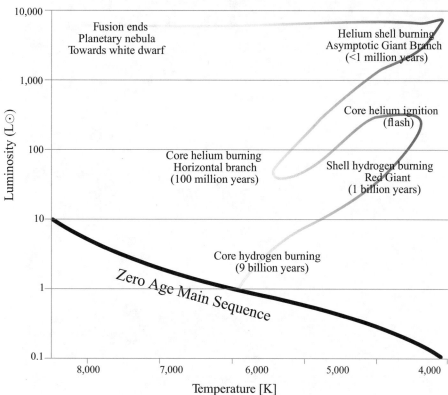

Figure 9.6: The evolutionary track of a star like the Sun, showing the zero age main sequence (ZAMS). The specific path shown is only schematic, and not from precise calculations as in Figure 9.5. (Graphic by Szczureq—Own work, CC BY-SA 4.0.)

The expanding outer layers are briefly illuminated by ultraviolet light from the contracting, extremely hot core. And this causes the gas to glow, forming a *planetary nebula*. Such an object has absolutely nothing to do with planets, except in a roundabout way. They were first identified as a type of object by William Herschel. Their superficial resemblance in a telescope to the planet Uranus (which Hershel discovered) led him to coin the phrase. See Figure 9.7 for one of the most famous examples, the Ring Nebula. It is visible in a small telescope as a faint smokey ring.

Planetary nebulae, although tiny compared to the vast Orion nebula, usually have beautiful symmetric shapes. Their full glory can be seen only if photographed with large high-resolution

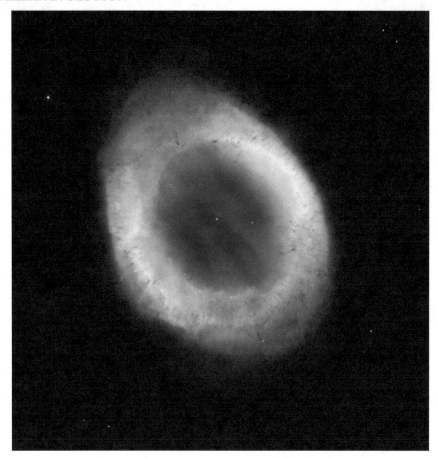

Figure 9.7: The Ring Nebula, M 57. This is one of the most famous examples of a planetary nebual. At its center is the hot contracting core of what used to be a red giant star; it will eventually contract to a white dwarf. (Image by The Hubble Heritage Team (AURA/STScI/NASA). Public Domain.)

telescopes such as the Hubble Space telescope. Many images can be found simply by performing an Internet image search on "planetary nebula."

9.6   LOWER-MAIN-SEQUENCE STARS

The stars with the lowest masses, way to the lower right on the main sequence, have the simplest fate. When they run out of hydrogen, they simply contract to a white dwarf and cool off; there is no red giant stage. But the estimated main sequence lifetime of such a star approaches a *trillion* years. As we have seen in Section 5.4.3, the universe is only 13.8 billion years old. And so no such star has yet had time to do this! All of the stars on the lower main sequence that have ever formed are still hissing away, slowly fusing hydrogen to helium, shining dimly compared to other stars. One such star, Proxima Centauri, is the closest star besides the Sun; yet it is far too dim to see without a telescope.

9.7   UPPER-MAIN-SEQUENCE STARS

The initial mass of a star decides its eventual fate. And the larger the mass, the faster everything happens. The most massive stars stay on the main sequence for only a million years or so; the Sun's main sequence lifetime is many thousands of times longer than this. But the changes that upper-main-sequence stars undergo are different in kind as well as degree.

Figure 9.5 shows evolutionary tracks for a variety of stellar masses, from only $0.1\,M_\odot$– $60\,M_\odot$. The figure labels the masses along the ZAMS, where the stars begin their visible lives. While on the main sequence, all of these stars change very little, drifting roughly perpendicular to the ZAMS. When they run out of hydrogen in their cores, they all change rapidly; one can see this rightward edge of the main sequence by the little kink in many of the evolutionary tracks.

The tracks for all of these stars, except for those very low on the main sequence, bring them to the upper right, either to the giant or supergiant region of the H-R diagram. That is, after leaving the main sequence they all initially (and relatively quickly) get cooler, larger, and brighter. But upper-main-sequence stars do this more quickly and to greater extremes.

Notice that the upper-main-sequence stars stay roughly the same luminosity for most of their visible lives. They start out hot and bright, and then after leaving the main sequence, become cooler, but also larger by just about the right amount to stay at roughly the same enormous luminosity. These very-massive stars are quite rare, because all stages of their lives happen so quickly. It is hard to catch one in the act of existing! But their enormous luminosities make them visible at vast distances, and so there are many examples among the common naked-eye stars.

9.8   STELLAR EXPLOSIONS

Stars on the upper main sequence die not with a whimper, but a bang. After every possible source of nuclear fusion energy is exhausted, the core of the star undergoes a catastrophic gravitational collapse, leading to an enormous explosion called a *type II supernova*. This explosion ejects much of the outer part of the star at speeds of tens of thousands of kilometers per second. Briefly, a supernova explosion can emit light equivalent to several billion Suns, making that single star as bright as a medium-sized galaxy.

These events are rare in any one galaxy, and in a spiral such as the Milky Way, they occur mainly in the disk, where dust obstructs our view of all but nearby ones. In 1054 AD, a supernova was recorded (mostly by Chinese astronomers) that was easily visible in the day time. The Crab Nebula is seen in telescopes at this same location today; see Figure 9.9.

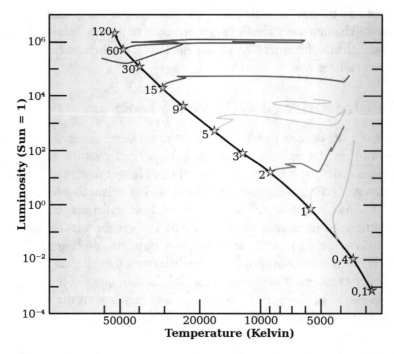

Figure 9.8: Evolutionary tracks for many different masses of stars. They all start at the zero-age main sequence (ZAMS), changing slowly and very little while on the main sequence. Then most undergo rapid changes to become red giants or supergiants. The very lowest mass stars do not become giants. (Graphic credit: Lithopsian, CC BY-SA 3.0.)

The glowing gas in this nebula is illuminated by shock waves, produced by fast-moving gas—just as one would expect for such a *supernova remnant*. The speed of the gas can be measured with the Doppler effect. A spectrum of the gas shows spectral features, and some are *both* redshifted and blueshifted. This means some of the gas is moving toward us, while other parts of the nebula are moving away from us. This is just as one would expect for a rapidly expanding shell of gas. Furthermore, photographs taken decades apart show the nebula literally getting larger. Extrapolating this expansion backward in time agrees with the date of 1054 AD for the explosion.

Most type II supernovae are discovered not in the Milky Way, but rather in other spiral (or irregular) galaxies. From literally one night to the next, a new star—comparable in brightness to the entire galaxy—appears in the galaxy's disk. Over many months, it fades away. Spectral analysis clearly shows high-velocity gas.

Figure 9.9: The Crab Nebula, what is left over from a supernova explosion in 1054. A neutron star lies at its center. (Image credit: NASA, ESA, J. Hester and A. Loll (Arizona State University), Public Domain.)

9.9   STAR CLUSTERS AND ISOCHRONES

By studying many individual stars, we can look for particular stages in the evolution of stars of different masses. We can then compare these observations to our theoretical predictions, illustrated for example by the evolutionary tracks of Figure 9.5. The results agree favorably.

But these are only snapshots in the lives of random stars, all of different ages and masses. It is difficult to untangle from these disconnected observations what stage in its life a particular star occupies. Differences in the abundance of "metals" adds to this complexity. Fortunately, there are a *lot* of stars to study! But even more important are the existence of star clusters. For a star cluster is a place where many stars—of different masses—formed at roughly the same time out of the same cloud of gas and dust. *And so the stars in a star cluster, although they have a range of masses, are all nearly the same age and chemical composition.*

This means that Nature has performed for us what are, in effect, controlled experiments in stellar evolution. A given star cluster shows us the progress along the evolutionary tracks of stars up to a particular point in time, *for many different masses at once.* Study a different star cluster and you will find the same thing, but it will likely show the results after a *different* amount of time has passed. Thus, taken as a whole, the star clusters show us snapshots in the evolution of stars.

If we plot the stars of a particular star cluster on the H-R diagram, we have what is in essence an *isochrone*—a diagram of constant time. We can then calculate *theoretical isochrones* using our theoretical models of the physics of stellar evolution, and compare them to the H-R digrams of known star clusters. To the extent that this provides self-consistent results, we can accept our theory, and then use it to determine the *age* of the star cluster.

Since clusters are groups of stars that are all the same distance from Earth, differences in apparent brightness correspond directly to differences in luminosity. Figure 9.10 shows the observed H-R diagram for two different open star clusters, M 67 and NGC 188, with the stars from each cluster plotted with a different color. Both clusters show the lower main sequence, but for each the upper main sequence is missing. That is because those stars, with much shorter main-sequence lifetimes, have long ago ended their visible lives. There is a portion of the main sequence that seems to be bending off to the right. This is called the *turnoff point* and it tells us that those stars are just now leaving the main sequence. There are also a small number of red giants on the upper right of the diagram.

The location of the turnoff point is an important clue to the age of the star cluster; its age is equal to the main-sequence lifetime of stars that are just now leaving the main sequence. If the turnoff point is higher up on the main sequence, the star cluster is younger; stars higher on the main sequence have a *shorter* main sequence lifetime. And so it is clear from this comparison that M 67 is younger than NGC 188.

9.10   WHAT REMAINS

All stars that start out with a mass greater than about $0.3 \, M_\odot$ eventually become red giants or supergiants. This is followed by a brief period of mass loss—much of the outer layers of the star are ejected away. For stars with a range of masses at the middle of the main sequence, this ejection is comparatively gentle, and briefly forms a beautiful planetary nebula. For upper-main-sequence stars, it is a violent supernova explosion.

But in all of these cases, the core of the star is thought to remain, and it will be some kind of small compact object—perhaps difficult to detect—that can simply cool off over time. These are the ultimate fates of stars, and I list them below in order of increasing *initial mass of the original star*:

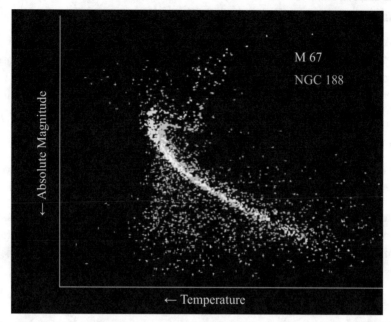

Figure 9.10: H-R diagrams for two star clusters, shown in two different colors. The lower main sequence is evident for both, as is the turn-off point. Since the turn-off point for M 67 is higher on the main sequence than that for NGC 188, M 67 is a slightly younger star cluster than NGC 188. (Graphic created by User:Worldtraveller, CC BY-SA 3.0.)

- **M less than 0.0125 M$_\odot$**

  Hot Jupiter: If the initial mass is too low, then the contracting ball of gas will never become hot enough inside to initiate any kind of nuclear fusion. It will be heated by gravitational contraction alone. Eventually it will contract, cool, and liquefy and become an object something more like the planet Jupiter than a star.

- **M greater than 0.0125 M$_\odot$ but less than 0.08 M$_\odot$**

  Brown dwarf: In this range of masses, the central core becomes hot enough to initiate some minor fusion reactions, but such a star will not be able to stablize and become a main sequence star. It will glow very dimly in the infrared.

- **M greater than 0.08 M$_\odot$ but less than 0.3 M$_\odot$**

  Helium white dwarf: These stars never become red giants, and after leaving the main sequence contract directly to a white dwarf made of almost pure helium. But this has not happened yet, because these stars have main sequence lifetimes longer than the age of the universe so far.

- **M greater than 0.3 M$_\odot$ but less than roughly 10 M$_\odot$**

  White dwarf made of carbon, neon, and oxygen, as well as helium: Stars with this range of initial masses become red giants or supergiants, and then go through a brief planetary nebula phase that ejects much of their mass. What is left is a white dwarf with a more complex composition than for the lowest mass stars. Higher temperatures in the cores of these stars has resulted in fusion of heavier elements.

- **M greater than about 10 M$_\odot$**

  Either neutron star or black hole: These most massive of stars end in a supernova explosion. What remains after the explosion clears is either a neutron star or a black hole.

The ultimate fate of a star depends on its initial mass, but there are also mass restrictions on the final product, which is often much less mass than the original star; many stars lose much of their initial mass in either the planetary nebula phase, or in a supernova explosion. Below I list the ranges of possible mass for these end products themselves.

1. **White dwarfs: M greater than 0.08 M$_\odot$ but less than 1.4 M$_\odot$**

   A white dwarf cannot be less massive than 0.08 M$_\odot$, because it would not have enough self-gravity to squeeze it to the high densities that characterize a white dwarf. But a white dwarf cannot be *more* than 1.4 M$_\odot$. This is the so-called *Chandrasekhar limit*. If one were to try to make a white dwarf with a greater mass than this, it would collapse and become a neutron star. Stars up to perhaps 10 solar masses eventually become white dwarfs, but they get rid of most of this mass in the planetary-nebula phase.

2. **Neutron stars: M greater than 1.4 M$_\odot$ but less than about 3 M$_\odot$**

   A neutron star cannot have a mass less than the Chandrasekhar limit, simply because its gravitational contraction would stop at the white dwarf stage. A neutron star can withstand more weight than can a white dwarf, but there is a limit. The calculation is less certain, but if one tried to make a neutron star with a mass more than roughly 3 M$_\odot$, it would instead collapse to a black hole.

3. **Black holes: M greater than about 3 M$_\odot$ but less than about 10 M$_\odot$**

   A black hole cannot form if the contracting mass is less than about 3 M$_\odot$, simply because the collapse would stop and become a neutron star instead. The upper limit is less certain. The most massive stars possible are roughly 100 M$_\odot$, but it is unknown just how much of that mass is ejected in the supernova explosion.

   Note that we are talking here about black holes formed as the end products of the evolution of stars. The *supermassive black holes* found in the centers of galaxies—with millions of times the mass of the Sun—are a different matter.

The most massive stars can make either a black hole or a neutron star, but I have not said which do one and which do the other. The reason is that it is not known with any clarity. All of the stars with initial masses greater than about $10\,M_\odot$ undergo supernova explosions, but it is difficult to calculate exactly what percentage of the mass of the star is lost in the explosion. And that is the key point; it is the mass of the contracting core left over that decides whether it will be a neutron star or a black hole.

It has, however, become increasingly clear that while neutron stars are relatively common, black holes formed by collapsing stars are rather rare. This could mean that only the very most massive (and thus rarest) stars eventually form black holes. But it could also mean that only a very narrow *range* of initial masses eventually lead to a black hole.

9.11   NUCLEOSYNTHESIS AND EVOLUTION OF THE ISM

Nuclear fusion in the cores of stars makes heavier elements out of lighter ones. In a supernova explosion, these fusion reactions generate all of the elements on the periodic table, as they are blasted into space to mix with, and become part of, the interstellar medium (ISM). This process of creating heavier elements from lighter elements by fusion in the cores of stars is called *nucleosynthesis*.

Recall from Part II of *The Big Picture* that the Big Bang produced mostly hydrogen and helium, with only minuscule traces of everything else—what astronomers refer to as "metals." And so the presence of what Earth—and we—are made of, is due to nucleosynthesis. The universe would be a far less interesting place without it, if hydrogen and helium were the only two elements. Helium does not combine chemically with anything, and so there would be only individual helium atoms, individual hydrogen atoms, and H_2 molecules. No bobolinks,[1] barley malt distillates, or ill-tempered felines in such a universe!

Since stars form from the ISM wherever it is dense enough, the process of nucleosynthesis continuously alters these parts of our Galaxy. In particular, the *metallicity* gradually increases over time, as successive generations of stars form, generate heavy elements in their cores, and blast these synthesized metals out into space to rejoin the ISM. And so a region of the ISM that shows a high metallicity betrays a billions-of-years history of continuous star formation.

---

[1]There are, unfortunately, fewer of these every year in the lovely universe we do have.

CHAPTER 10

The Evolution of Galaxies

A galaxy is not a "thing." Rather, it is the sum of all its parts—individual stars, star clusters, gas, and dust—all held together by gravity. And a galaxy is a *dynamic* structure; each piece must move according to the sum of gravitational forces acting upon it. But also, the individual stars, clusters, and nebulae within a galaxy evolve with time according to their own local conditions. And so for a galaxy, both the individual pieces, and how those pieces are arranged in space must change as time passes. A casual view through a telescope of a spiral galaxy similar to our Milky Way shows only the overall effect of this history of change—the individual stars, for example, are usually too faint to see.

10.1   FORMATION AND EVOLUTION OF THE MILKY WAY

The structure of our Galaxy provides clues to its formation. In particular, there are separate *populations* of stars. A full description is more complex, but the two most extreme opposites have the following properties.

Thin disk population (extreme Population I):

- Consists of hot upper-main-sequence stars, young open clusters of stars, and clouds of gas and dust.
- Travel on circular orbits narrowly confined to the disk of the Galaxy, all orbiting in the same sense.
- Relatively high metallicity.

Halo population (extreme Population II):

- Consists of older individual stars and the globular clusters.
- Orbits of high inclination to the disk of the Galaxy, and in essentially random directions.
- Relatively low metallicity.

The most obvious conclusion from these observations is that the halo population of stars formed long before the thin disk population. The still-present gas and dust in the disk of the Galaxy allows for continuous star formation. And so a recently formed open star cluster is the result of many generations of star formation and supernova explosions. This process *enriches* the interstellar medium (ISM) with metals. And so the ongoing processes of stellar evolution and nucleosynthesis creates a stellar nursery that is enriched in metals.

The halo seems to be devoid of the relatively dense concentrations of gas and dust necessary for the formation of stars. And so perhaps the globular clusters and the individual stars in the halo formed very early on in the history of the Galaxy. And so the gas and dust they formed from was much nearer to primordial—the mix of nearly pure hydrogen and helium provided by the Big Bang itself.

In the next section I outline two scenarios for the formation of the Milky Way; an excellent discussion can be found in Binney and Merrifield [1998, Section 10.7]; my description largely follows theirs.

10.1.1   TOP-DOWN SCENARIO

The overall shape of a disk galaxy like the Milky Way bears a striking resemblance to the solar system, but on a vastly larger scale. Both have a disk-like structure, the parts of which orbit together in the same direction, confined to a narrow plane. And they both have a concentration of matter in the center—the central bulge in the case of the Milky Way and the Sun in the case of the solar system. Furthermore, each has a spheroidal population of objects that orbit in random directions—the globular clusters and halo stars for our Milky Way and the Oort cloud comets for the solar system.

These similarities naturally lead to the idea that the Milky Way may have formed in a manner similar to how the solar system is thought to have formed. A large cloud of gas contracts due to self-gravity, but a small amount of initial rotation leads to the formation of a thin disk, most-concentrated in the center.

The idea is that the globular clusters and the first generations of stars formed *during* that initial gravitational contraction, and they retain the orbits of that initial motion—plunging toward the center of the Galaxy.

Note that when a cloud of gas contracts due to its self-gravity, the gas is compressed, and this is what allows it to form a disk; gas does not pass through gas. But already-formed stars *do* pass easily through gas. And so the halo stars can pass through the disk of the Galaxy in their plunging orbits.

After the original gas cloud had collapsed into a disk, stars could no longer form in the halo of the Galaxy. And so stars that formed later formed in the disk itself, and they bear the circular orbits of the disk gas they formed from. This *top-down scenario* was first proposed in detail by Eggen et al. [1962].

10.1.2   BOTTOM-UP SCENARIO

The top-down scenario of Eggen et al. [1962] is simple and attractive, but it has been called into question in light of more recent observations. It is unrealistic, at least in detail, for two reasons. First, galaxies formed in an expanding universe, and this greatly changes the dynamics of a contracting proto-galactic cloud. And second, it fails to take into account the fact that galaxies often join together in *mergers*.

Certain features of our Galaxy make better sense if *bottom-up* processes are taken into account instead of an overall top-down gravitational contraction from a rotating proto-galactic cloud of primordial gas. Modern attempts to explain the formation of the Milky Way and other spiral galaxies often use *N-body simulations*. Gravitational forces are calculated, and the motions determined, individually for a myriad of mass points—stars if you will. After a small step in the motions of the mass points, the forces are recalculated and the process begins again. Large, high-speed supercomputers are required for some of the most advanced N-body simulations; we consider them further in Chapter 17, in the context of the overall large-scale structure of the universe.

N-body simulations show that small systems of stars are constantly merging together to form new and bigger ones. This challenges the notion of a single, isolated galaxy that retains its identity while slowly evolving. Galaxies are not only the sum of their parts, they are the sum of their histories. But this does not mean the top-down scenario of Eggen et al. [1962] is incorrect in its entirety.

10.2   REFERENCES

James Binney and Michael Merrifield. *Galactic Astronomy*. Princeton University Press, 1998. 140

O. J. Eggen, D. Lynden-Bell, and A. R. Sandage. Evidence from the motions of old stars that the Galaxy collapsed. *The Astrophysical Journal*, 136:748, November 1962. DOI: 10.1086/147433 140, 141

PART IV

Process

CHAPTER 11

Fields

11.1 NEWTON'S GRAVITY

The Law of Universal Gravitation, formulated by Isaac Newton in the late 17th century, was the first description of a universal physical law in the modern sense. It built upon the foundation of Galileo and Kepler, a generation before.

Newtonian gravitation has as its foundation the notion that every point of *mass* in the universe experiences an attractive force toward every other point of mass. This force, F_G, has a strength that is proportional to the product of the two masses, and inversely proportional to the square of the distance between them. An intrinsic symmetry is apparent: each mass feels the same force toward the other. But since the force is attractive, this means these two forces are opposite in direction; it is a *mutual* attraction.

If we denote the two masses as m_1 and m_2, and the distance between them as d, we can write this law as follows:

$$F_G = G\frac{m_1 m_2}{d^2}. \tag{11.1}$$

Here G represents a constant of proportionality, the value of which depends upon the details of our particular system of units of measurement, combined with the quantitative strength of gravity. In our SI system of units, we measure the masses in kilograms (kg), the distance in meters (m), and the force in *Newtons* (N).[1] And so the numerical value of G represents the force of gravity in Newtons between two 1-kg masses placed one meter distant from each other.

We can solve Equation (11.1) for G:

$$G = \frac{d^2 F_G}{m_1 m_2}. \tag{11.2}$$

And so to measure G we need only place two 1-kg masses 1 m apart, and then measure the gravitational force each feels toward the other. Newton did not perform this obvious experiment, because it is equally obvious that this force must be extremely small—far too small to be measured in Newton's time. Otherwise, our world would be a very different place; the stuff of our everyday existence would want to gravitate together into one lump. The *constant of universal gravitation*, G, was first measured in the late 1700s—some seven decades after Isaac Newton's death—by Henry Cavendish, using a sensitive torsion balance. Modern measurements yield G

---

[1] A kilogram of mass here on the surface of Earth weighs about 9.8 N.

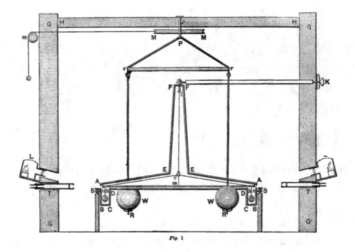

Figure 11.1: An illustration of the torsion balance used by Henry Cavendish to measure the constant of universal gravitation. (Graphic Public Domain.)

to be a tiny quantity indeed:

$$G = 6.674 \times 10^{-11} \, \mathrm{N \, m^2 \, kg^{-2}}. \tag{11.3}$$

See Figure 11.1 for an illustration of Cavendish's apparatus.

The very small measured value for G means that gravitation is a relatively weak force, measurable on our everyday scale only with the most sensitive of instruments. Electrical forces, for example, are overwhelmingly larger, all else being equal. The gravitational force between two 1-kg masses placed one meter apart is only 6.674×10^{-11} N, equivalent to the weight here on Earth's surface of a microscopic droplet of water only 0.024 mm ($24 \, \mu$m) across. This is about the weight of a microscopic protozoan such as *Trypanosoma cruzi* (which causes Chagas disease) or *Entamoeba histolytica*.

Gravity is the force that most concerns us as citizens of Earth. But this is only because we happen to be relatively close to an enormous quantity of mass—the 5.972×10^{24} kg of mostly rock and metal that makes up Earth. For astronomers, it is gravity that holds it all together and has the greatest effect on the motions of large bodies in the universe, even though it is the intrinsically weakest of the physical forces, all else being equal. The reason is that all else is *not* equal; gravitation is the only one of the fundamental forces that is both long-range and *only* attractive. The much-stronger electrical force, for example, can either attract or repel, and so it tends to cancel out at large distances.

11.1.1 THE CLASSICAL GRAVITATIONAL FIELD

Equation (11.1), taken at face value, asserts that one object exerts an *instantaneous* influence (a gravitational force) upon another, even if the two are separated by a vast expanse of the vacuum of space. But it is possible to look at Newton's gravity in a different way; we replace the action at a distance with the *local* action of the *gravitational field*.

Let us return to our example of two point masses, m_1 and m_2, placed a distance d from each other. From the point of view of Newton, we say that m_1 exerts a gravitational force, at a distance, on m_2. But from the perspective of the gravitational field, we say something subtly different: *the presence of m_1 causes there to be a gravitational field, $\vec{g}$, that extends throughout space.* We can think of this field as a property of space itself. And so we have the gravitational field, $\vec{g}$, due to the presence of m_1. Our field theory must then tell us how to calculate this field. For this simple case of the gravitational field due to a single point mass, the answer is very simple. The field points everywhere toward m_1, and it has this magnitude, for a point in space located a distance d from m_1:

$$g = G\frac{m_1}{d^2}. \tag{11.4}$$

But this is only half the story; we must also say what the field *does*. And the answer to this is even simpler:

The gravitational field, $\vec{g}$, at a given point in space exerts a force on any mass, m, that is placed there. And that force is given by $\vec{F}_G = m\vec{g}$. The gravitational field thus has dimensions of force per unit mass (SI units of $N\,kg^{-1}$).

The little arrows are there to remind us that both the gravitational field and the gravitational force are in the same direction.

We can now consider what happens to m_2 from the perspective of the gravitational field. The mass m_2 knows nothing about m_1; it only knows that locally—where it is—there is a gravitational field, $\vec{g}$. This local gravitational field exerts a gravitational force on m_2 in the direction of $\vec{g}$, and with magnitude $F_G = m_2 g$. But we have already seen that $\vec{g}$ points toward m_1, and so too does $\vec{F}_G$. And we also know the magnitude of $\vec{g}$ from Equation (11.4). Thus, we have

$$F_g = G\frac{m_1}{d^2}m_2. \tag{11.5}$$

The result is the same as what we already knew; m_2 experiences a force toward m_1, and given by Equation (11.1).

If the local field conception of a gravitational field always produces the same result as Newton's original action-at-a-distance, then why bother with fields at all? Let us consider a different example. Suppose that we attach some kind of meter to m_2 that allows us to directly measure the gravitational force acting upon it. And then suppose we quickly move m_1 a bit closer to m_2, thus making d smaller. Obviously, the force meter would then read a higher value; both the action-at-a-distance and local-field conceptions agree.

But does the increase in F_G occur *instantaneously*? Or is there some sort of time delay? If gravity acts instantaneously at a distance, it means we could use the gravitational force to send messages across the vastness of space, with no time delay at all. Clearly, this would violate special relativity, which implies that no influence can travel faster than light (see Section 5.1).

Newton's action-at-a-distance conception of gravity implies, in effect, that the gravitational interaction acts *instantly*, no matter the distance of space. Considering gravity instead as the *local* action of a field provides a way out of this dilemma. For we could imagine that the when we move m_1, the resulting change in the gravitational field—a property of space itself—propagates outward from m_1 at a finite speed in some kind of wavelike disturbance. Object m_2 would then react only when that disturbance in the gravitational field reached its location. This basic concept is called a *gravitational wave*, and there is no possibility for such a phenomenon in Newton's theory of gravity. If we were to attempt to modify Newtonian gravity so as to allow for gravitational waves, a field-theory perspective would be essential.

Einstein's theory of gravity, completed a decade after his published paper on special relativity, is a field theory. But it is not a force field like our field-theory formulation of Newton's gravity. Rather, Einstein's gravitational field is the effect matter has on the very geometry of space and time itself. We consider Einstein's theory of gravity more fully in Section 11.2.

11.1.2   GRAVITY AND SPHERES

Much of the motivation for the Cavendish experiment was not directly to measure the constant of proportionality, G, in Newton's law of gravity. Rather, Cavendish wanted to measure the mass (more specifically the density) of *Earth*. What is the connection? Consider Figure 11.2.

A small object of known mass—a beaver[2] in this example—sits on the surface of Earth. It experiences the downward force of gravity we call *weight*, easily measurable with a balance or scales. This "downward" force is, by observation, toward the center of Earth. This should be unsurprising; the symmetry of the law of gravity and the arrangement of masses dictates that this is the only possible direction. But what is the magnitude of this force? Equation (11.1), in and of itself, does not provide the answer, for it applies to two *point-like* masses.

To apply Equation (11.1) to this example, one would need to break Earth up into little tiny pieces, each of which is small enough to be considered point-like. Then the law of gravity could be applied to each piece with Chapter 2—noting that even if the masses are chosen to be equal, the distances are not. But also, one would need to take the quite-different *directions* of these myriad tiny forces into account when adding them up. This process—breaking a problem up into tiny pieces, calculating something for each piece, and then adding up all of the results—is part of the foundation of *calculus*. These mathematical techniques did not exit in Newton's day, so he had to invent them. We now call it *integration*. For this particular case, although the process is complex, the answer is simple:

---

[2]Not drawn to scale.

Figure 11.2: The gravitational force between two spheres acts as if all of the mass of each sphere were concentrated at its center. (Earth image: NASA/Apollo 17 crew, Public Domain. Beaver image by Steve, Washington, D.C., CC BY-SA 2.0.)

> The gravitational force between *spherically symmetric* distributions of mass acts as if all of the mass were concentrated at their centers.

With this mathematical result, the way forward is clear; the force between the beaver and Earth *can* be calculated with Equation (11.1). We simply use for d the distance between the center of the beaver[3] and the *center* of Earth. If we assume this is not actually a giant space beaver, and so it is minuscule compared to Earth, then this distance is, essentially, the radius of Earth. Thus, we have:

$$F_G = G \frac{m_{\text{beaver}} M_\oplus}{R_\oplus^2}. \qquad (11.6)$$

---

[3] Assume a spherical beaver.

We can rearrange this to solve for the mass of Earth:

$$M_\oplus = \frac{F_G R_\oplus^2}{m_{beaver} G}. \tag{11.7}$$

And so if we measure the force, F_G, on the beaver of known mass, and we know the radius of Earth, then the knowledge of the gravitational constant, G, leaves only the mass of Earth, which we can solve for and then calculate. Thus, the Cavendish measurement of the constant of universal gravitation was equivalent to determining the mass of Earth.[4]

This measurement of the mass of Earth also allowed for the calculation of its average *density*—its mass divided by its volume. Comparing this to densities of materials on Earth showed that it is about $5500\,kg\,m^{-2}$. This is more than that of typical rocks, but less than that of metals such as iron, strongly hinting that Earth is made of a combination of both materials.

We can also consider this result from the point of view of the gravitational *field*; the gravitational field due to a spherical distribution of matter is the same as if all of its mass were concentrated at its center. And so we can easily calculate the magnitude of the gravitational field near the surface of Earth:

$$g = G\frac{M_\oplus}{R_\oplus^2}. \tag{11.8}$$

The direction of $\vec{g}$ is toward the center of Earth, and this is of course our meaning of the word "down." If we put in the accepted values for the gravitational constant and the mass and radius of Earth, we find that, to two significant figures, $g = 9.8\,N\,kg^{-1}$. The strength of the gravitational field can be measured directly; it is equal to the acceleration of any freely-falling body. And so simply drop an object near the surface of Earth, and measure its acceleration, a. If the effects of other influences such as air resistance are eliminated, you will find that $a = 9.8\,m\,s^{-2}$; a meter per second squared of acceleration is the same thing as a newton per kilogram of gravitational field.

11.1.3 GRAVITY AND ORBITS

The measurement of G allows for the determination of the mass and density of the Moon as well. It was known since Newton that there is a relation between the sum of the masses of two astronomical bodies and their orbits about each other:

$$m_1 + m_2 = \frac{4\pi^2}{G}\left(\frac{a^3}{p^2}\right), \tag{11.9}$$

where p is the *period* of the orbit (the time for one complete orbit), and a is half the long axis of the ellipse-shaped orbit; this is simply the radius if the orbit is circular. See Figure 11.3.

Equation (11.9) is Newton's version of *Kepler's Third Law*. It means that if we can determine both the size and period of an orbit, the sum of the masses of the two bodies can be calculated.

[4]It was also necessary to find an appropriately cooperative beaver.

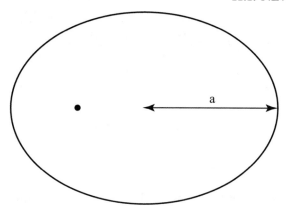

Figure 11.3: The periodic orbit of a small body about a much more massive one is, in general, elliptical in shape, with the larger body offset from the center. The relevant size of the orbit in Equation (11.9) is represented by a, the *semimajor axis* (half the longest dimension). A circular orbit is simply a special case of an ellipse with zero eccentricity.

For the case of the solar system, we are used to thinking of a planet (Jupiter, for example) orbiting about the Sun. But really, they both orbit about each other—or rather, about a point in space along the line joining them, known as the *center of mass* of the orbit. The center of mass is proportionally closer to the object of *greater* mass; it is in effect the balance point between them. For the case of the Sun and Jupiter, this point is deep inside the Sun. And so Jupiter makes a large orbit, while the Sun merely wobbles a bit.

Figure 11.4 shows a diagram of the orbit of a hypothetical binary star system. The left side depicts the *relative orbit*—the orbit of the less-massive star using the more-massive star as the point of reference. The right side shows the true orbits for both stars about their common center of mass.

With the measurement of the period and semimajor axis from the relative orbit on the left side of Figure 11.4, Equation (11.9) allows us to calculate the *sum* of the two masses. If the true orbit is also known, as in the right side of Figure 11.4, we can measure the ratio of the distance each star is from the orbit's center of mass. If we know both the sum of the masses and their ratio, it requires only simple algebra to calculate each mass separately.

Variations on this basic process are the foundation for our determinations of the masses of nearly everything in the universe. We measure both the arrangement of astronomical bodies in space and the motions that arise due to the mutual gravitational forces between them. To use this process to calculate masses requires that we know the distances between objects in real *physical units* such as meters, light years, or astronomical units. It is not enough to merely know how far apart two objects *appear* to be, as seen from Earth. And so in many cases, the *distance* to the orbiting bodies must also be measured in order to determine their masses.

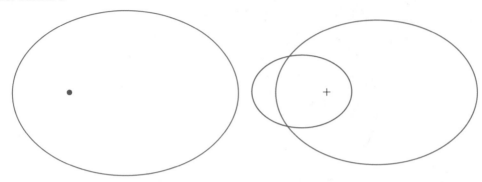

Figure 11.4: **Left:** The relative orbit (shown in blue) of one star about another (shown in red). **Right:** Both stars actually make separate orbits about a common center of mass (represented by the cross) that lies always along the line joining the two orbiting bodies. The center of mass is closer to the more massive body, in inverse proportion to the ratio of their masses.

11.1.4   WHAT IS MASS?

Newton's law of universal gravitation, combined with his laws of motion, imply that there are two separate definitions of "mass." By this I mean *operational definitions*; we define the physical quantity by describing the very procedure use to measure it. In the context of Newton's laws, there are two such procedures for measuring the mass of an object.

1. **Inertia:** Kick it, and see how much your foot hurts.

2. **Gravity:** Place it on a balance here on the surface of Earth, and see how much it weighs.

These two procedures for measuring mass seem so different and unrelated to each other that it would seem unjustified to even use the same word—"mass"—to signify their results. But a simple experiment shows that both definitions give the same result. Simply drop different objects near the surface of Earth. If they all fall with the same acceleration, these two very-different definitions of mass give—mysteriously from Newton's point of view—the same answer.

Newton's second law of motion relates external forces acting on an object to the acceleration that results. The two are proportional to each other, and it is the mass of the object that connects the two. We might call this way of measuring mass the *inertial mass*, m_i, of the object:

$$m_i = \frac{\Sigma \vec{F}}{\vec{a}},$$

(11.10)

where $\Sigma \vec{F}$ is the net external force acting upon the object and a is its acceleration, its rate of change of velocity. The little arrows over the variables are there to emphasize that both of these quantities have directions as well as magnitudes. Mass, on the other hand, has only a magnitude.

But if gravity is the only force acting upon the object, then Newton's law of universal gravitation proclaims that:

$$\Sigma \vec{\mathbf{F}} = m_g \vec{\mathbf{g}}, \tag{11.11}$$

where $\vec{\mathbf{g}}$ is the local strength of the gravitational field where the object is dropped. This is mass as measured by gravity, and so I have used m_g to represent the *gravitational mass* of the object.

If we combine Equations (11.10) and (11.11) for the case of an object falling under the influence of gravity alone, we have:

$$\vec{\mathbf{a}} = \frac{m_g}{m_i}\vec{\mathbf{g}}. \tag{11.12}$$

Equation (11.12) implies that if $m_g = m_i$, then $\vec{\mathbf{a}} = \vec{\mathbf{g}}$ and the acceleration is simply equal to the local value of the gravitational field, *regardless of what mass is placed there*. And so all objects would experience the same acceleration in a given gravitational field, if they are acted upon only by gravity.

We can thus test whether or not $m_g = m_i$ by performing a simple experiment; drop different objects in a gravitational field and measure whether or not their accelerations are indeed equal. Galileo was the first to describe this experiment, allegedly dropping objects off the leaning tower of Pisa.[5] Since Newton's time, the experiment has been performed repeatedly, to higher and higher levels of precision. And no one has ever convincingly discovered evidence for any difference between gravitational and inertial mass. This experimental result—that inertial and gravitational mass are equivalent to each other even though they seem to have very different definitions—is one way to state Einstein's *equivalence principal*, a cornerstone of his general theory of relativity we discuss in Section 11.2.

11.2 EINSTEIN'S GRAVITATIONAL FIELD

Instead of somehow "tweaking" Newton's law of gravity to include a speed-of-light time delay, Einstein started over. He took Minkowski's cue that special relativity could be interpreted as a *geometric* theory (see Section 5.1.4), and spent the next ten years developing an elegant new theory of gravity based on geometry. It is called *general relativity* (GR), and its foundation was published in 1915. Some of the most basic principals of GR are listed below.

- **The equivalence principal.** There is no distinction between gravitational mass and inertial mass—there is only mass (and it's equivalent energy). This means that *the motion of every material body is affected the same, in a given gravitational field.* The equivalence principal also means that, *when looking at only a very small region of space and time*, gravity is indistinguishable from an ordinary acceleration.

[5]Interestingly, it is not clear whether Galileo actually performed the experiment at all. His argument, foreshadowing Einstein in many ways, was rather that it *should* be true that all objects would fall the same [Ferris, 1988, pp. 83–94].

- **Gravity is not seen as a force in GR.** Material bodies move under the influence of gravity the same way they move when under no force at all—they move in straight lines. But in GR, bodies move in straight lines not in ordinary three-dimensional space, but in *four-dimensional spacetime*. Such a straight line in spacetime is called a *geodesic*.

- **The very geometry of spacetime is altered by the presence of mass to be non-Euclidean** (see Section 2.1.1). The presence of mass alters both the geometry of space and the rate at which time passes. This means the geometry of four-dimensional spacetime is altered by the presence of mass.

- Because mass alters the geometry of spacetime, **mass changes the meaning of what is a straight line in space and time**. And so particles move in different paths through space than they would if no mass were present.

- **Light beams travel on *null geodesics*.** A null geodesic occupies zero time—the light beam is everywhere along its path at once. But it travels in a straight line through ordinary (but curved) space alone. And so light beams can be used to trace out the non-Euclidean three dimensional space, curved by the presence of mass.

And so the presence of mass tells space and time how to "curve"—to assume a geometry different from the ordinary Euclidean geometry we know and love. Matter then moves in what is a straight line through this altered four-dimensional geometry—and thus follows a path that is different than it would have if no mass had been present.

Earth does not *feel* a force of gravity from the Sun. Rather it is simply moving in a straight line through whatever space and time it happens upon. But that spacetime has been altered by the presence of the enormous bit of mass we call the Sun. And so there is no spooky "action at a distance;" all action is local.

11.2.1   GRAVITATIONAL LENSES

Because light rays trace out the local curvature of space, the presence of a large amount of mass—from a galaxy or cluster of galaxies for example—can act as a sort of lens. More distant light rays bend around the concentration of matter, and this means rays of light that would have missed us can be deflected in our direction.

See Figure 11.5 for the most extreme example—an Einstein ring. Rays from many angles have been deflected to us, stretching the image of a distant galaxy into a thin ring. Far more often, the alignment of the gravitational lens is not so perfect.

Gravitational lenses such as that shown in Figure 11.5 can be used to probe the curvature of space on the large scale. This provides another observational constraint on cosmological models that assume the curvature of the universe is flat on the large scale.

Figure 11.5: An example of the most rare type of gravitational lens—an Einstein ring. The thin ring of light is a galaxy behind the one at its center. The curvature of space has bent light from the more-distant galaxy around the nearer one, allowing us to see it in many directions at once. (Image credit: ESA/Hubble and NASA, CC BY-SA 4.0.)

11.2.2   NEUTRON STARS AND BLACK HOLES

A neutron star is an extremely compact object; an enormous amount of mass (roughly twice the mass of the Sun) is contained within a tiny volume (a sphere the size of a small city). This means the curvature of spacetime nearby is not at all subtle.

This affect of gravity on spacetime can be measured directly in a few cases, such as the double pulsar PSR J0737-3039. A pulsar is a rapidly-rotating neutron star, that emits beams of radio waves somewhat like a light house. If we happen to be in the path of the beam, it flashes toward Earth every rotation of the neutron star. Thus, we see a regular flash of these radio waves.

Because of the conservation of angular momentum, pulsars rotate very fast, with periods of only seconds or even tiny fractions of a second. The pulsars in PSR J0737-3039 rotate with periods of 2.773 seconds and 22.699 *milli*seconds, respectively. The short orbital period of a pulsar makes it, in effect, a clock. And so we can use the pulses to monitor the difference between the pulsar's rate of time passage and our own.

Detailed analyses of the orbits of this binary pulsar *require* GR; if GR is ignored in the calculations, then the predicted orbit disagrees with what we observe. GR on the other hand— including the production of gravitational waves—gives answers that agree with observation.

If the mass of a neutron star were to be compressed to only slightly smaller volume, it would form a *black hole*—an object so dense that the extreme curvature of space prevents even a light beam from escaping it.

If mass is confined within a certain radius such that even light cannot escape the gravity, then a black hole is formed. This radius is known as the *Schwarzschild radius,* and it sets the effective size of a black hole. And so we imagine a sphere with a radius equal to the Schwarzschild radius. The surface of this sphere is called the *event horizon*—anything that passes through it cannot return.

The Schwarzschild radius, R_S, of a black hole depends only upon its mass, M, and it has a surprisingly simple formula:

$$R_S = \frac{2GM}{c^2},\tag{11.13}$$

where G is the gravitational constant and c is the speed of light. If we express M in units of the mass of the Sun, the Schwarzschild radius is approximately 3 km for every solar mass.

Black holes greatly alter the spacetime nearby, but at large distances their gravity is not significantly different from what it would be if the same amount of mass were not compact and in the form of a black hole.

The black holes that remain from the collapse of massive stars are very difficult to detect; it is practically impossible for a solitary black hole. But in a close binary star system, the companion star may transfer matter to the black hole, and this produces X-rays that are detectable. There are several such examples known.

But by far the most easily detectable black holes are the *supermassive black holes* found at the centers of many large galaxies. Figure 11.6 shows the first successful "photograph" of a supermassive black hole—in the center of the giant elliptical galaxy M 87.

This remarkable image was released in April, 2019 by the Event Horizon Telescope team. The dark hole in the center of the image is larger than the Schwarzschild radius of the black hole. It represents the edge of the *ergosphere*—a region around a black hole where light beams would essentially orbit the concentration of mass. The mass calculated for this supermassive black hole is about 6.5 *billion* times the mass of the Sun.

11.2.3 DARK MATTER AND DARK ENERGY

The nature of *dark matter* (see Section 5.4.5) is consistent with a type of matter that behaves, so far as gravity is concerned, the same as any other matter. It is its interactions with *other* known forces (electricity, magnetism, and the forces within the nucleus of an atom) that are unusual. But so-called *dark energy* (see Section 5.4.6) is another kettle of fish. Dark energy is the name for the recent observation that the universe is accelerating its expansion, rather than slowing down as one would expect from "normal" gravity. The cause of this acceleration is unknown, but it has a deep connection, historically if nothing else, to Einstein's GR.

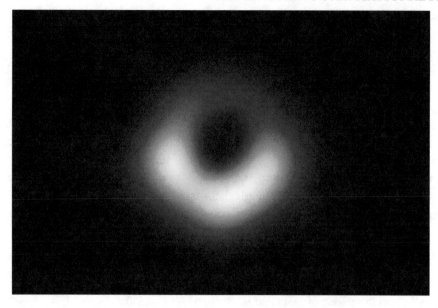

Figure 11.6: An image of the ergosphere of the supermassive black hole in the center of the giant elliptical galaxy M 87, made with the Event Horizon Telescope. (Image credit: EHT Collaboration, CC BY 4.0.)

Shortly after Einstein published his first papers on GR, he realized that it had cosmological implications. His equations implied that the universe could not be static—it must either expand or contract. This was 1915, and the expansion of the universe would not be discovered by Edwin Hubble for decades.

Einstein believed, wrongly, that the universe as a whole must be static. And so in 1916 he proposed to modify his equations with a *cosmological constant*. It was a fudge factor. The term, signified by the Greek letter Λ (lambda), allowed for what amounts to a repulsive effect of space itself at very large distances. This could, for the universe as a whole, balance the overall attractive effect of "ordinary" gravity. On the small scale (in the laboratory or even the solar system), it would be essentially undetectable.

When Hubble discovered the expansion of the universe, Einstein disavowed the cosmological constant, calling it his biggest blunder. But with the development of quantum physics over the next few decades, other physicists were not so sure. Quantum physics *does* predict that space itself should have some kind of energy associated with it—and GR says that energy should have gravitational effects. But even a quick quantum-mechanical calculation gives an answer that is not only wrong, it is absurd.

And so for many decades, no one knew what to do with Einstein's cosmological constant. There are some theoretical reasons for expecting it to exist, but actual calculations with existing theory give absurd results. And so there was not much for a cosmologist to do but assume the cosmological constant to be zero, unless new evidence were to arise that suggests a different value.

This all changed in the late 1990s when observations indicated that the expansion of the universe is *accelerating*, rather than decelerating. This observed effect is well fit by Einstein's cosmological constant.

The term "dark energy" is more popular than "cosmological constant" to describe this observed acceleration. We have no coherent theoretical underpinning for this effect on the expansion of the universe, and so the "dark" part of the term emphasizes the mystery. And it suggests a linguistic kinship with the not-quite-as, but still very mysterious *dark matter*. Still, its historical origins in Einstein's cosmological constant are preserved in the symbol Λ, which cosmologists use to represent the observed effect of this mysterious property of the universe.

11.3   THE ELECTRIC AND MAGNETIC FIELDS

The phenomena of electricity and magnetism have been known since antiquity. Magnetic rocks occur naturally, and the effects of so-called *static electricity* are easily observed. But it was not until the late 19th century that a complete understanding developed of the wide variety of electrical and magnetic effects. The culmination of these insights is the first true field theory, formalized in the 1860s and 1870s by James Clerk Maxwell, building on the work of many others.

Maxwell's field theory explains all known electrical and magnetic phenomena with four equations—known as Maxwell's equations. We will neither solve nor use Maxwell's equations in this book! But even absent an understanding of the mathematics, there is much one can learn simply by looking at them; they are the four Equations (11.14)–(11.17):

$$\nabla \cdot \vec{E} = \frac{\rho}{\epsilon_0} \tag{11.14}$$

$$\nabla \cdot \vec{B} = 0 \tag{11.15}$$

$$\nabla \times \vec{E} = -\frac{\partial \vec{B}}{\partial t} \tag{11.16}$$

$$\nabla \times \vec{B} = \mu_0 \vec{J} + \epsilon_0 \mu_0 \frac{\partial \vec{E}}{\partial t}. \tag{11.17}$$

Each of the four Maxwell equations refers to one—or both—of the *electric field* (symbolized by $\vec{E}$) and the *magnetic field* ($\vec{B}$). There are some other symbols as well.

- The symbols ϵ_0 and μ_0 are constants that, roughly speaking, scale the measured strengths of electricity and magnetism, respectively.

- The symbol ρ (called *charge density*) is a measure of the concentration of electric charge at any given point in space.

- The symbol $\vec{\mathbf{J}}$ (called *current density*) represents the rate and direction of *flow* of charge through any give region of space.

- The symbol t represents time.

All of the other symbols in Equations (11.14)–(11.17) are purely mathematical in nature, and do not refer to any physical phenomena *per se*.

There are some other things to note. First, the electric and magnetic fields (as well as current density) are given symbols in boldface type, topped by an arrow. This is to signify that they are *vector quantities*—their directions matter as well as their strengths. And so $\vec{\mathbf{E}}$ and $\vec{\mathbf{B}}$ represent *vector fields*.

In Maxwell's theory, these fields permeate all of space; in fact, we consider them to be *part of the very properties of space itself*. Accordingly, every point in space has a particular value—consisting of both a magnitude *and* a direction—of both $\vec{\mathbf{E}}$ and $\vec{\mathbf{B}}$. How does one then determine the values of $\vec{\mathbf{E}}$ and $\vec{\mathbf{B}}$ for some region of space? Solve Equations (11.14)–(11.17)!

We will not solve Maxwell's equations here, but we will consider further what they *mean*. But even so, this is clearly only part of the story. Maxwell's equations allow one to calculate the electric and magnetic fields at any point in space. But so what? We also need to say what these fields *do*. What are the physical consequences if the electric and magnetic fields have one value instead of another?

These questions bring to the fore one of the hallmarks of any field theory. Rather than describing directly how one thing affects another, a field theory breaks the interaction into two separate questions.

1. What are the physical *causes* of the field? What is it that affects the value of the field at some point in space, and how do we calculate it?

2. What are the physical *consequences* of the field? If there is a field at some point in space, what happens?

What makes a field theory such as Maxwell's equations complex and subtle is that the answer to each question is affected by the answer to the other. For electricity and magnetism, Maxwell's equations are the answer to the first question. We will consider the answer to the second question in Section 11.3.3, and then further consider the meaning of Maxwell's equations in Section 11.3.1. Finally, in Section 12.2, we will consider an important and revolutionary consequence—unexpected at the time—of Maxwell's equations.

11.3.1 WHAT THE FIELDS DO

What are the direct physical consequences of electric and magnetic fields in space? The answer lies in the mathematical description of what is known as the *Lorentz force*, $\vec{\mathbf{F}}_L$:

$$\vec{\mathbf{F}}_L \;=\; q\vec{\mathbf{E}} + q\left(\vec{\mathbf{v}} \times \vec{\mathbf{B}}\right). \tag{11.18}$$

The Lorentz force is exerted on any particle with *electric charge*, q, located in the presence of $\vec{E}$ or $\vec{B}$, or both. And so in rough terms, the answer is this: electric and magnetic fields in space exert *forces* on any electric charges located there. These *electric forces* and *magnetic forces* will then alter the motions of those charged particles.

The right-hand side of Equation (11.18) has two parts that are added together to calculate the total effect on the charged particle:

- $q\vec{E}$, the *electric force* and

- $q(\vec{v} \times \vec{B})$, the *magnetic force*.

Both of these forces, considered individually, are proportional to the electric charge in question. So particles with a greater electric charge experience a greater force, all else being equal.

But the electric and magnetic forces act in quite different ways. The electric force is simpler; its strength is simply proportional to—and in the same direction as—the electric field, $\vec{E}$.[6] This means, for example, that electric fields can easily be used to accelerate charged particles to high velocities. Simply arrange for a region of space to have a uniform electric field throughout, and a charged particle placed there will feel a constant electric force from that field. Thus, the particle will accelerate faster and faster. A consequence is that electric fields can *do mechanical work* on charged particles, and so increase (or decrease) their *kinetic energy*. Thus, there is an energy associated with electric fields.

The magnetic force, on the other hand, only acts upon charged particles that are *moving*. The symbol $\vec{v}$ in Equation (11.18) represents the *velocity* of the charged particle in question—its speed, combined with its direction of travel. And so a magnetic field has no effect at all on a *motionless* charged particle.

But the action of magnetic fields is even more intriguing. The magnetic force relates only to the part of the charged particle's motion that is *perpendicular* to $\vec{B}$. Only motion crosswise to the magnetic field counts. And furthermore, the magnetic force itself is always directed perpendicular to *both* the magnetic field and the motion of the charged particle.

Because of this cross-wise nature of the magnetic forces, they cannot be used to change the *speed* of a charged particle. Instead, they change the direction. A force that is always crosswise to one's motion is just what is needed to cause a swirly motion. And so charged particles tend to swirl around magnetic fields rather than move along them. This also means the magnetic force cannot *directly* alter the kinetic energy of a charged particle.

11.3.2   WHAT IS CHARGE?

The existence of the Lorentz force—a force that is exerted only on particles that have the property of electric charge—begs an important question: what exactly *is* electric charge? At one level,

---

[6]If the charge is negative, the electric force acts in the direction opposite the electric field.

we can say simply that "it is what it is." Charge is a property of certain fundamental particles—the electron and proton being the most familiar examples—such that they interact with each other according to the rules set down by Maxwell's equations.

But is there a deeper explanation? Why is charge quantized? Why does the smallest unit of charge have the particular value that it does? Why are there two kinds of charge? Why do, for example, electrons and protons have charge, but neutrons do not? Why is charge conserved?

These questions cannot be answered from within the confines of Maxwell's 19th-century theory of electricity and magnetism. But modern physics does provide at least some hints of answers to some of these questions. The conservation of charge, for example, can be seen to arise as a consequence of a particular *symmetry* of nature (see Section 14.2.2). And some sense has been made—at least at the level of categorization—of the charges and other properties of the many different fundamental particles (see Section 14.3). But many of the deep questions about the fundamental nature of charge are beyond current understanding (see, for example, Penrose [2004, p. 66]).

11.3.3   MAXWELL'S EQUATIONS AND THE CAUSES OF $\vec{E}$ AND $\vec{B}$

The four Maxwell equations, (11.14)–(11.17), represent two equations each for the two vector fields, $\vec{E}$ and $\vec{B}$. Notice that there are two versions of the left-hand side, repeated for each field.

- Equations (11.14) and (11.15) begin with "$\nabla\cdot$." This is called a *divergence*, and it refers to a vector field that diverges away from—or converges toward—some point in space.

- Equations (11.16) and (11.17) begin with "$\nabla\times$." This is called a *curl*, and it refers to a vector field that swirls around a point in space.

And so Maxwell's four equations answer two questions about each of the two fields, $\vec{E}$ and $\vec{B}$.

1. What causes the field to diverge from or converge toward points in space?

2. What causes the field to swirl around points in space?

From a mathematical perspective, to answer both of these questions is to say all there is to say about a vector field. For the electric and magnetic fields, the answers are on the right-hand sides of Equations (11.14)–(11.17). An intriguing thickening of this plot is the fact that two of the Equations—(11.16) and (11.17)—include $\vec{E}$ and $\vec{B}$ in their right-hand side; thus the answer is, in effect, part of the very question. Furthermore, $\vec{E}$ is in the right-hand side of the equation for $\vec{B}$, and vice versa. Thus, the two fields are *coupled*; one cannot really know one, without also knowing the other. In that sense, the electric and magnetic fields $\vec{E}$ and $\vec{B}$ are but two individual sides of a complex and subtle *electromagnetism*.

And so let us use words to relate the right-hand sides of Maxwell's equations to their left-hand sides, in the same order as they appear in Equations (11.14)–(11.17).

Gauss's Law

Equation (11.14) is called *Gauss's Law*. It says that a diverging electric field is caused by the very presence of electric charge in space. This is, for example, the origin of the simple force of static electricity. Put an electric charge on the table, and electric field diverges away from it in all directions.[7] Put another charge next to it, and Equation (11.18) for the Lorentz force says the field will exert a force on that second charge, in the same direction as the electric field.[8] Thus, it is as if the one charge puts a force of repulsion (or attraction) on the other.

Krauss's Law

Equation (11.15) has no particular historical name, but it is clearly mathematically similar to Gauss's law, except that it applies to $\vec{B}$ instead of $\vec{E}$. Many years ago one of my physics students—named Jeff Krauss—gave me permission to use his name for this equation. And so "Krauss's Law" says that a diverging magnetic field is caused by—nothing at all! Magnetic fields *never* diverge away from (or converge upon) points in space. This is sometimes taken to be equivalent to the statement that there is no such thing as "magnetic charge."

Faraday's Law

Equation (11.16) is called *Faraday's Law*, and it says that a swirly electric field is caused by a *magnetic* field that changes with time. The magnet sitting motionless on the door of your refrigerator has no effect on the electric field. But move the magnet back and forth in front of the door and it will cause electric fields to swirl around in space. These electric fields will then, according to the Lorentz force, accelerate electrons in the conducting metal door along those swirly $\vec{E}$ fields. This causes a swirling *electric current* in the metal. These are called *induced currents*, and they can occur in an isolated electric circuit disconnected from any external source of power.

The Ampère-Maxwell Law

Equation (11.17) is the most complex of the four, and the second term on the right-hand side is Maxwell's addition to what had been known as Ampère's Law. This last piece of the puzzle transformed the four equations into a mathematically coherent field theory. And so although the rest were discovered by others (Gauss, Faraday, and Ampère in particular), we honor the set, taken as a whole, with Maxwell's name.

The Ampère-Maxwell law says that there are two ways to make a swirly *magnetic* field. But first, "Krauss's Law" has already told us that there is no such thing as a diverging magnetic field. And so the only kind of magnetic field that exists is one that swirls back upon itself to form a closed loop of magnetism.

---

[7]If it is a negative charge, then $\vec{E}$ will instead converge toward it.

[8]If it is a negative charge, the force is opposite the direction of $\vec{E}$.

The first term on the right-hand side of Equation (11.17) contains the current density, $\vec{J}$. This is the Ampère's Law part, and it says that electric charge that moves through a region of space will cause a magnetic field to swirl around that motion. And so it is well known that magnetic fields swirl around a wire that carries an electric current. It is only *moving* charge that has this effect; a stationary charge will cause, in light of Gauss's Law, only an *electric* field.

Maxwell's added term in Equation (11.17) states that there is a second way to make a swirling magnetic field—with an *electric* field that changes with time. If this seems reminiscent of Faraday's Law, it should. Apart from the scaling factors and lack of a negative sign, it appears just like Faraday's law, *but with the roles of $\vec{E}$ and $\vec{B}$ reversed.*

11.4   REFERENCES

T. Ferris. *Coming of Age in the Milky Way*. William Morrow and Company, New York, 1988. 153

Roger Penrose. *The Road to Reality: A Complete Guide to the Laws of the Universe*. Vintage Books, 2004. 161

C H A P T E R   12

Waves

12.1 THE NATURE OF WAVES

Wave phenomena[1] allow the transfer of energy and momentum from one place to another without actual stuff having to make the trip. Move your hand up and down in the water at one end of the bathtub, and eventually the rubber duck at the other end bobs up and down too. Yell, "Hurry up!" to your roommate in the bathroom and very quickly their eardrums vibrate, even though no air from your mouth traveled to their ear.

This calls attention to the fact that a wave by its very nature is extended in both space and time. It is a pattern spread out in space that moves with time. And even at a given location, the wave changes as time passes. And so there is no real meaning to assigning a precise location and time to a wave. We can talk about what the wave does at a particular place and time, but we need to consider all of the other places and times in order to describe any particular wave. In this way, it is very unlike, say, a stone moving through space, which has a much more precisely definable position at any specific time.

A wave in its purest form has four basic attributes: speed, amplitude, wavelength, and frequency. We consider each of these in turn.

12.1.1 AMPLITUDE, SPEED, WAVELENGTH, AND FREQUENCY

One might point out that if no stuff actually makes the trip when a wave moves from one place to another, what do we really mean then by the "speed" of a wave? If the wave is very complex, then this question may have a complicated answer. But for a simple wave, the answer is easy:

A wave is a repeating pattern in space that moves as time passes. The speed of the wave is the speed at which this *pattern* moves.

Let us consider a wave traveling down a string, in the way that a wave travels down a garden hose when one end is waved up and down. For now, let us imagine the string to be infinitely long, so we can ignore the interesting complication of waves reflecting off the ends. If one were to take a flash picture of the wave, freezing it in time, a repeating pattern in space would be evident. One can represent this pattern in the form of a graph, as in Figure 12.1.

The meaning of the amplitude, A, of the wave is clear from the diagram. But there is another equally important measure. The *distance* over which the wave completes one repetition

[1]Some of the material in this chapter appeared, in a somewhat different form, in Part II of *The Physics and Art of Photography, Volume 1: Geometry and the Nature of Light* [Beaver, 2018a].

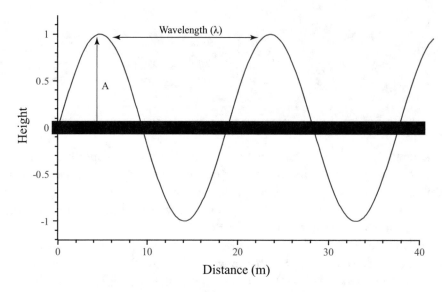

Figure 12.1: Wavelength (λ) and amplitude (A) of a wave. Notice that the horizontal axis of the graph is length. The wavelength is the *distance* over which the wave repeats itself, *at a particular point in time*.

is called the *wavelength* of the wave. By historical convention, we use the Greek letter λ (lambda) as our symbol for wavelength.

We can also represent a wave in terms of changes in time, rather than in space. Figure 12.2 looks, at first glance, the same as Figure 12.1; but note the horizontal scale. Instead of showing the wave at different points in space (at a given instant of time), Figure 12.2 shows the wave as time passes (but at some particular point in space). And so the *period, T,* of the wave is the *time* required for one repetition of the wave to pass a particular point in space, as the wave goes by. In SI units, the period is measured in seconds; for most electromagnetic waves of concern to astronomy, this is a tiny fraction of a second. More commonly we refer to the *frequency, f ,* the reciprocal of the period:

$$f = \frac{1}{T}.$$ (12.1)

The frequency then would be measured in inverse seconds, labeled s^{-1}, or Hertz (Hz). It is the number of wavelengths that go by per second.

While the period for electromagnetic waves is typically a very small number, the frequency is thus typically a very large number. There are two reasons for this. First, the wavelengths of visible light are very tiny. Second, the speed of light is very large.

Clearly, the faster is the speed of the wave, the more wavelengths would go by in one second. But also, the shorter the wavelength, the more that would go by per second, for a given

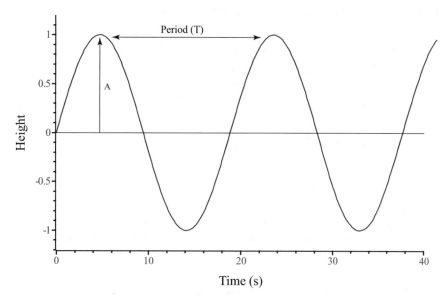

Figure 12.2: Period (T) and amplitude (A) of a wave. Notice that the horizontal axis is time. The period is the *time* over which the wave repeats itself *at a particular point in space*. The *frequency* (f) of the wave is the reciprocal of the period.

speed. Thus, we have a relation between the frequency, f, the wavelength, λ, and the speed, v, of a wave:

$$v = f\lambda. \tag{12.2}$$

12.2 LIGHT: ELECTROMAGNETIC WAVES

The last two Maxwell equations—Faraday's Law (11.16) and the Ampère–Maxwell Law (11.17)—are in many ways the most intriguing. Let us consider in particular a simple but special case—the vacuum of empty space itself. If there is nothing there at all, then certainly there is no moving charge, and thus no charge density, $\vec{J}$. And so these two equations become simply:

$$\nabla \times \vec{E} = -\frac{\partial \vec{B}}{\partial t} \tag{12.3}$$

$$\nabla \times \vec{B} = \epsilon_0 \mu_0 \frac{\partial \vec{E}}{\partial t}. \tag{12.4}$$

These two equations, taken together, say that a magnetic field, $\vec{B}$, that changes with time causes there to be a swirling electric field, $\vec{E}$. But also, the opposite is true: an electric field, $\vec{E}$, that changes with time causes there to be a swirling magnetic field, $\vec{B}$.

I can easily make a changing electric field; simply jiggle an electric charge back and forth. Gauss's law says that electric field diverges away from the charge. And so if the charge is moving back and forth, the electric field at any given point in the empty space around it must also be changing with time. But in that empty region of space, the changing $\vec{E}$ will cause, according to Equation (12.4), a magnetic field $\vec{B}$. And in this particular case the magnetic field induced will *also be changing with time.*

You can see where this is going. The induced and changing magnetic field $\vec{B}$ would then produce in the space around it—by Equation (12.3)—a swirling electric field $\vec{E}$, which by Equation (12.4) would produce a changing magnetic field, and so on.

Maxwell was the first to realize the profound implications of Equations (12.3) and (12.4) taken together, even for the empty vacuum of space. Once one produces a changing electric or magnetic field (by, for example, jiggling a charged particle back and forth), the two fields inevitably act upon each other to produce a changing pattern of electricity and magnetism that moves rapidly through space—an *electromagnetic wave.*

In hindsight, it is only a matter of sophomore-level physics to demonstrate that Equations (12.3) and (12.4) can be combined mathematically to form the *wave equation.* In the process, the speed of this wave can be calculated directly from the scale factors ϵ_0 and μ_0—already known in Maxwell's time from simple laboratory experiments of electricity and magnetism. The answer is this: the speed of the wave of electricity and magnetism is $2.998 \times 10^8 \, \mathrm{m\,s^{-1}}$. I hope that by now you recognize this speed as c, the speed of light.

And thus light *is* an electromagnetic wave—a rapidly changing pattern of electric and magnetic field that can move through even the vacuum of space. An electromagnetic wave can do this because the electric and magnetic fields are properties of space itself, and so there need be nothing there at all. But if on the other hand there *is* something there—a charged particle for example—the electromagnetic wave will do stuff to it as it passes by. For the fields exert forces on charged particles.

And so, move a charged particle back and forth *over here*, and an electromagnetic wave will radiate outward through space at the speed of light. The traveling electromagnetic wave will then exert forces on a charged particle *over there*, and so cause stuff to happen. This is the meaning of light.

12.2.1   THE ELECTROMAGNETIC SPECTRUM

We use the symbol c to represent the speed of light—the speed of an electromagnetic wave traveling in a vacuum. And so for light, Equation (12.2) becomes:

$$c = f\lambda. \tag{12.5}$$

Because the speed of light, c, is a constant, we can see there is a simple relationship between wavelength and frequency for electromagnetic waves. Thus, any given wavelength corresponds to a particular frequency, and vice versa. We can rearrange Equation (12.5) as follows:

$$f = \frac{c}{\lambda} \tag{12.6}$$

$$\lambda = \frac{c}{f}. \tag{12.7}$$

These are *reciprocal relations*; if frequency is larger, then wavelength is smaller, and vice versa. It also means that, for light, we can choose either frequency or wavelength for our description. If given one, the other can be easily calculated.

Different wavelengths (or frequencies) of electromagnetic waves interact with matter in different ways. Since both the absorption and emission of light are examples of such interactions, one would need different strategies to *produce* light of vastly different wavelengths. Likewise, different methods are required to *detect* light of very different wavelengths.

The human eye is only sensitive to the very narrow range of wavelengths between about 0.4–0.7 millionths of a meter, and so it is this range of wavelengths that defines what we call *visible light*.

The electromagnetic wave nature of light was unknown until the late 1800s. And it wasn't until the 20th century that most forms of electromagnetic waves were finally identified. Some types had previously been detected, but it wasn't recognized until much later that they were just different wavelengths of electromagnetic waves. And so different ranges of wavelength of electromagnetic waves have different names, in part for historical reasons.

Table 12.1 shows the ranges of possible wavelengths, along with their customary names. Taken together, this is called the *electromagnetic spectrum*. Keep in mind that the ranges of wavelengths or frequencies are only approximate; the boundaries are fuzzy and overlap each other. The names really come from the different ways in which we produce or detect them, and that has changed over the years as technology has changed.

And so let us very briefly consider each of these basic parts in turn. I will start at the long-wavelength bottom of the list; this may seem strange, but remember that *long* wavelength is the same as *low* frequency.

- **Radio waves** are made by moving an electrical current back and forth in a wire, and radio waves induce currents to oscillate back and forth in wires they pass through.

- **Microwaves** can be thought of as very high-frequency radio waves. They can sometimes be made in the same fashion, but other processes (besides electronic circuits) are also used to produce them. Because of their shorter wavelength, they can be focused with special mirrors and more easily guided along paths.

- **Infrared light** is usually produced in ways similar to visible light, but the wavelengths are too long for the human eye to detect it.

Table 12.1: The electromagnetic spectrum

Name	Typical λ(m)	Typical Size	f (Hz)
Gamma Ray	$< 1 \times 10^{-11}$	Atomic nucleus	$> 3 \times 10^{19}$
X-Ray	$1 \times 10^{-11} - 3 \times 10^{-8}$	Atom	$1 \times 10^{16} - 3 \times 10^{19}$
Ultraviolet	$1 \times 10^{-8} - 4 \times 10^{-7}$	Virus	$7.5 \times 10^{14} - 3 \times 10^{16}$
Visible Light	$4 \times 10^{-7} - 7 \times 10^{-7}$	Bacteria	$4.3 \times 10^{14} - 7.5 \times 10^{14}$
Infrared	$7 \times 10^{-7} - 1 \times 10^{-3}$	Protozoa	$3 \times 10^{11} - 4.3 \times 10^{14}$
Microwaves	$1 \times 10^{-4} - 0.1$	Person	$3 \times 10^{9} - 3 \times 10^{12}$
Radio	> 0.1	Building	$< 3 \times 10^{9}$

- **Visible Light** is the name for the narrow range of wavelengths (about 400–700 nm) to which the human eye is sensitive. Different wavelengths of visible light produce different color sensations; violet for short wavelengths and red for long wavelengths, with blue, green, yellow, and orange in between.

- **Ultraviolet light** also is often produced in ways that are similar to visible light, but the wavelengths are too short to be detected by our eyes.

- **X-rays** have wavelengths similar to the size of individual atoms, and so they usually interact with matter in ways that involve individual atoms on a one-on-one basis.

- **Gamma rays**, with their enormous penetrating power, interact directly with the tiny nuclei of atoms. And thus they are associated most strongly with nuclear reactions.

12.2.2   LIGHT AND ITS SPECTRUM

There are many ways in which light can be created by matter. But any particular method for producing light is essentially an interaction between light and matter, and the wavelength (or frequency) of the light is of crucial importance.

For example, to make electromagnetic waves of a frequency of 10^6 Hz, simply use an electronic oscillator circuit to make a current go back and forth in a wire at that frequency. A frequency of 10^6 Hz is in the radio part of the electromagnetic spectrum, and so this is essentially the basis of a radio transmitter. But for visible light with frequencies several powers of ten higher, this strategy simply will not work. Instead one can, for example, heat up a tungsten wire to a high temperature, and the individual tungsten atoms will vibrate at high frequencies, and visible (and infrared) light will be emitted.

Whenever light is created by matter, some wavelengths are made well while others are made poorly or not at all. Thus, to really describe a particular source of light, we need to describe

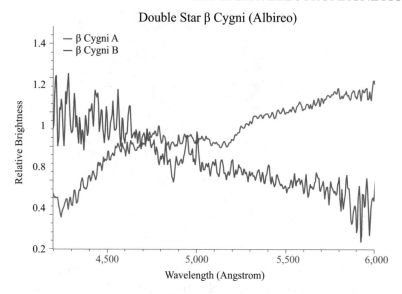

Figure 12.3: The spectra of two stars making up the binary star system β Cygni. Each colored line represents the spectrum of one of the stars. The fine-scale squiggles result from noise (random measurement uncertainty), but the overall trends in the spectra are accurate. The spectra span the range from 4,200–6,000 Å (420–600 nm), thus covering the range from blue-violet on the left to orange-red on the right (data from Beaver and Conger [2012]).

how much of each wavelength has been produced. Such a description is called a *spectrum* (plural, spectra) of the light source, and the most useful way to represent one is with a graph.

Figure 12.3 shows the spectra of the two stars that make up the binary star system Albireo. The horizontal axis of the graph is wavelength, with short wavelength on the left and long wavelength on the right. The axis goes from about 420 −600 nm, and this is visible light, within the range of wavelengths sensitive to the human eye. The spectrum of one of the stars is marked in red while the other in blue. Clearly, β Cygni A emits more long-wavelength (red) light than short-wavelength light (blue), while the opposite is true for β Cygni B.

In the case of β Cygni and other stars astrophysicists can determine many things about their physical natures by analyzing the spectra of the light they emit. We will use this idea of the spectrum of a light source many more times throughout this book; it is one of the most important tools for understanding light.

12.2.3   THERMAL RADIATION

The term *thermal radiation* refers to electromagnetic waves emitted due to the normal motions of atoms and molecules in some material. Whether it be liquid, solid, or gas, the atoms and molecules that make it up are constantly in motion. Although the individual particles are moving

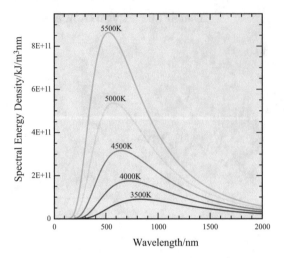

Figure 12.4: The blackbody spectra for several different temperatures. Blackbody radiation is the most ideal form of thermal radiation. The curve marked 5500 K is similar to the spectrum of the Sun. Note that higher-temperature objects are both brighter and radiate at predominately shorter wavelengths. (Graphic by 4C, CC BY-SA 3.0.)

with a wide range of energies (called kinetic energy), there is in many circumstances a meaningful average. This average of the kinetic energies of the individual particles in a material is directly related to what we call *temperature*. Thus, the atoms and molecules making up a hot potato are moving, on average, with greater energy than are those of a frozen potato.

These motions of the individual atoms and molecules generate electromagnetic waves, and the waves produced have a wide range of frequencies, just as the atoms and molecules have a wide range of energies. Thus, a *spectrum* of frequencies (or wavelengths) is produced. If all of the individual atoms and molecules are in perfect balance with each other—and with the light they produce—then a particular kind of spectrum is produced called a *blackbody spectrum*. That perfect balance of matter and light, called *thermodynamic equilibrium*,[2] ensures that *the details of the blackbody spectrum depend only on temperature.*

Figure 12.4 shows five different blackbody spectra, with temperatures ranging between 3500 −5500 K, where "K" stands for Kelvin, our SI unit of absolute temperature. The Kelvin scale starts at absolute zero, and room temperature is roughly 300 K. The surface temperature of the Sun is 5770 K, so the temperatures represented in Figure 12.4 are very high indeed. Nearly all of the chemical elements are vaporized at 5000 K.

The graph is silent about what *type* of material made these spectra for a very simple reason—it doesn't matter. A 5000 K blackbody is a 5000 K blackbody, whether it is produced from calcium, carbon, cobalt, or vaporized chocolate-chip cookies.

[2]We will consider thermodynamic equilibrium in more detail in Chapter 13.

Table 12.2: Peak wavelengths for different temperature blackbody spectra

T(K)	λ_{max} (μm)	Part of E-M Spectrum	Example
2.7	1,100	Microwave	Cosmic microwave background
300	10	Thermal infrared	Room tempertaure objects
3,000	1	Near infrared	Light bulb filament
6,000	0.5	Visible light	Visible surface of Sun
10,000	0.3	Near ultraviolet	Visible surface of Vega

The overall shape of a blackbody spectrum is that of a broad hump, with a peak at some particular wavelength. There is a particular mathematical formula that describes the exact shape of a blackbody spectrum, but there are two basic features that are obvious from the graph alone.

1. **The Stefan–Boltzmann Law:** The blackbody spectrum's hump is bigger—which means more energy is emitted—if the temperature is higher. This is a huge effect, as the brightness is related to the total *area* under the graph. Notice how much larger is the area under the 5500 K, compared to that under the 3500 K graph. *The brightness of a blackbody scales with the fourth power of the temperature.* And so a doubling of the temperature increases the brightness by $2^4 = 16$ times.

2. **Wein's Displacement Law:** A *higher* temperature produces a blackbody spectrum that peaks at a *shorter* wavelength. This is described by a simple inverse proportionality. As an example, we can see from the graph that a 3000 K blackbody peaks at a bout 1 μm (1000 nm). And so a room-temperature 300 K blackbody, with one tenth the temperature, peaks at ten times that wavelength, or 10 μm, in a region of the spectrum known as the *thermal infrared*.

Table 12.2 shows the peak wavelength for different temperatures. A blackbody the temperature of the Sun, a bit less than 6000 K, peaks right in the middle of the visible spectrum. The part of the Sun's atmosphere that makes the visible surface of the Sun is not in perfect equilibrium (if it were, the light could not escape). And so the visible spectrum of the Sun is not quite the same as a blackbody—but it is a rough approximation of one, with the same overall shape. Notice that the light from an ordinary light bulb is on average of much longer wavelengths than that of the Sun.

12.2.4   THE COLORS OF STARS

The spectra of most stars is very roughly that of a blackbody, and so the perceived color of a star is strongly dependent on its temperature. The human eye is sensitive only in the range 400 − 700 nm, while a hot blackbody emits at a much broader range of wavelengths; see Figure 12.4. A cool star such as Betelgeuse, in Orion, has a temperature not very different from the light bulb filament listed in Table 12.2—about 3000 K. The visible surface of the Sun, on the other hand, is roughly twice that temperature, while the hottest stars are nearly 30,000 K.

A glance at Figure 12.4 shows that a blackbody with a temperature similar to that of the Sun emits most of its light at ultraviolet and infrared wavelengths, even though its spectrum peaks at about 500 nm, near the middle of the visible spectrum. This is simply because our eyes are sensitive to only a relatively narrow range of wavelengths.

Our eyes see only the part of the star's spectrum that happens to be in the narrow range of visible wavelengths. And so the perceived color of the star arises from the star's particular mix of these visible wavelengths only. For a star like the Sun, this light peaks in the yellow-green part of the visible spectrum, and so sunlight has more of these wavelengths than either violet or red. To the human eye, this appears essentially as white.[3]

For a cool star on the other hand, the spectrum peaks in the infrared. See the 3500 K curve in Figure 12.4; it cuts diagonally downward across the 400 − 700 nm visible spectrum from right to left. And so we see far more of the long-wavelength red light than the short-wavelength violet light. Our perception of this spectrum is a pale reddish. For the hottest stars, the opposite is true. Their spectra peak in the ultraviolet, and so we see far more violet light than red light. These stars thus appear pale bluish in color. See Figure 12.5 for a reasonably accurate rendition of these perceived-by-humans star colors, as generated by the star map software Cybersky (available at www.cybersky.com).

12.2.5   NON-THERMAL RADIATION

Thermal motion of atoms and molecules is not the only way to produce light, and so not all sources of light emit with a spectrum approximately like that of a blackbody. A laser pointer, for example, emits only one wavelength of light, not a broad range like a thermal source. And there is no simple connection between temperature and the wavelength of the laser. Non-thermal sources of light such as lasers and emission-line sources produce light at the atomic level. That is, each atom produces its own light, emitting individual photons. And so the particle-like nature of light must be considered to understand these sources. These are one-on-one interactions between light (in the form of individual photons) and matter (in the form of individual atoms). As such, the spectrum of the light produced *depends critically on the type of atoms emitting the light*. This is the opposite of the case of thermal radiation. *For purely thermal radiation, only the temperature matters*; the type of atom is irrelevant. Many real sources of light, when considered in more detail, involve a combination of both thermal and non-thermal emission.

---

[3]Human perception of white and nearly white colors is strongly dependent upon context. For more on this complex topic, see Beaver [2018b, Chap. 10 and Sec. 11.2].

Figure 12.5: An approximate reproduction of the visual appearance of the varieties of colors of stars, generated by the star map software Cybersky (available at www.cybersky.com), used with permission.

12.2.6   THE SPECTRUM OF AN EMISSION NEBULA

A good astronomical example of a non-thermal spectrum is that of the glowing clouds of gas we call *emission nebulae*. See Figure 12.6 for the spectrum of the planetary nebula NGC 6543. It shows only light at very precise single wavelengths.

The type of spectrum seen in Figure 12.6 is known in general as an *emission or bright-line spectrum*. Once can accomplish something similar simply by putting a low-density gas in an otherwise evacuated tube, and using electrodes to subject the gas to a very high voltage. The high voltage *ionizes* the gas; some of the electrons are stripped off the atoms.

The gas in an emission nebula such as NGC 6543 or the Orion Nebula is also low density and ionized—but it is ionized not by a high voltage but rather by the action of ultraviolet light from nearby hot stars. In the case of NGC 6543 it is the star at its center—the core of a former red giant now contracting to become a white dwarf. For the Orion Nebula, the Trapezium stars are the culprit. This process is called *photoionization*—literally, ionizing with light.

The specific spectral lines in an emission nebula spectrum depend most crucially upon the elements, ions or molecules that are present in the gas. Secondarily, the relative strengths of these lines can be used to infer physical properties such as temperature and density.

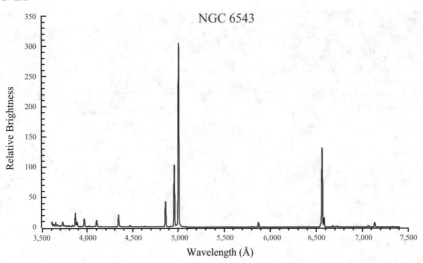

Figure 12.6: The spectrum of the planetary nebula NGC 6543, a low-density ionized gas, is an example of an emission spectrum. (Data by J. Montier, BAA.)

12.2.7   STELLAR SPECTRA

Stars have a more complex spectrum than either blackbodies or emission nebulae. The inside of a star is about as close to a blackbody spectrum as anything in nature. But we don't see that directly, because the interior of the star is opaque due to the extreme scattering of light by free electrons in the fully-ionized gas.

This blackbody spectrum finally escapes through the star's atmosphere, and it is greatly altered by these higher, cooler, and lower-density layers of gas as it passes through them. The gases *absorb* some of that light—but only at certain specific wavelengths.

This produces an overall continuous spectrum much like a blackbody, but with super-imposed dark lines or bands at the wavelengths of absorption. Such an *absorption or dark-line spectrum* is typical for stars. See Figure 12.7 for two examples.

The spectrum on the left in Figure 12.7 is of a star cooler than the Sun, while that on the right is considerably hotter. One can see that the peaks in these spectra are at wavelengths one would expect for cooler and hotter blackbodies in general; the spectrum of the hotter star peaks at a shorter wavelength.

But the patterns of absorption lines are completely different for the two stars. The very deep absorption lines in HD 109995 are due to hydrogen, while these hydrogen lines are very weak in HD 041636. Instead strong lines of "metals," such as ionized calcium, are present. This turns out to be almost entirely due to the different *temperatures* of the atmospheres of these stars; their chemical compositions are almost identical. Thus, the precise combinations of spectral lines

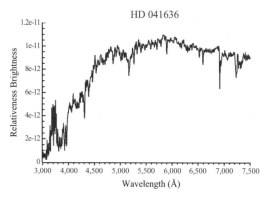
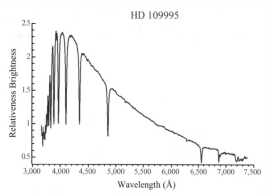

Figure 12.7: The absorption spectra of two different stars. Left: HD 041636 is a star that is cooler than the Sun. Its overall spectrum is brightest near the yellow-orange part of the visible spectrum. It shows weak absorption lines of hydrogen, but very strong absorption by ionized calcium. Right: HD 109995 is considerably hotter than the Sun, and so its spectrum peaks at near-ultraviolet wavelengths. It shows strong absorption lines of hydrogen. Both stars show absorption lines superimposed upon the overall spectrum. Even though the two stars show very different absorption lines, their chemical compositions are similar. (Data by T. Rodda, BAA.)

observed in the spectra of stars is a sensitive measure of temperature—more so than the overall color of the star.

That the absorption spectra of different stars could be arranged in a pattern—an ordered sequence of spectral types—was first understood in the early 1900s by Williamina Flemming, Antonia Maury and Annie J. Cannon, all working at Harvard College Observatory. The history of this development is complex and interesting, but the end result is our modern sequence of spectral types signified by the letters O, B, A, F, G, K, and M. It was Cecilia Payne, in the mid-1920s who first fit a temperature scale to this spectral sequence, applying then-new principals of statistical quantum physics, developed by the Bangladeshi astrophysicist Meghnad Saha, to stellar atmospheres.

12.2.8 BRIGHTNESS AND THE INVERSE SQUARE LAW

If a source of light is *isotropic*—it shines uniformly in every direction—then geometry provides a simple relation between how bright it is, how bright it appears, and its distance. The *inverse square law* states that the *apparent brightness* of a source of light is proportional to its *luminosity* divided by its distance squared.

Luminosity, L, is a measure of *power*, the amount of energy per second emitted by the source of light, and in our SI system it has units of watts. The apparent brightness, on the other hand, is a measure of the intensity, I, of light that we receive on Earth; it is an *energy flux*, and has units of watts per square meter.

If the source emits light isotropically, then at a distance, d, all of the emitted power must pass through a sphere of that radius. And so we must have:

$$I = \frac{L}{4\pi d^2},$$

(12.8)

where I have used the fact that the surface area, A, of a sphere of radius, d, is given by $A = 4\pi d^2$.

Note that this relation is for objects that emit their own light. And so Equation (12.8) is most useful for stars and galaxies. I have also assumed that the light emitted by the source has not been absorbed or scattered off in other directions on its way to Earth, as would be the case if the intervening space were filled with dust. We shall see that this factor is often important, and it leads to a more complex relationship between luminosity, intensity and distance.

For astronomical objects such as stars and galaxies, the luminosities and the intensities as seen from Earth vary by many powers of ten. For this reason, astronomers most often use a logarithmic scale, called the *magnitude scale*. The intensity as seen from Earth of the light from an astronomical object is described by an *apparent magnitude*, denoted by m. The luminosity is described by an *absolute magnitude*, denoted by M.

For historical reasons, the astronomical magnitude scale is set up not as a common logarithm based on powers of ten; rather, it is defined in the following way.

- For a common logarithm, a difference of one corresponds to a factor of ten. But the magnitude scale is instead defined such that *a difference of 5 magnitudes corresponds to a factor of 100*. This corresponds to a factor of approximately 2.512 in brightness for a difference of one magnitude; $2.512^5 \approx 100$.

- A *larger magnitude* corresponds to a *dimmer* object.

- An apparent magnitude of zero is, by convention, assigned to an apparent brightness roughly that of the bright star Vega. The brightest naked eye star, Sirius, has $m = -1.4$, while the faintest one can see with the naked eye is approximately $m = 6$.

- The absolute magnitude, M, of a star is defined to be the apparent magnitude it would have if it were at an agreed-upon fixed distance of exactly $10\,\mathrm{pc}$.

Given the definition of our magnitude scale and the properties of logarithms discussed in Section 1.1, it is easy to derive Equation (12.9) for the relation between the apparent magnitudes, m_1 and m_2, for two sources of light, and their intensities, I_1 and I_2, as measured in physical units such as $\mathrm{W\,m^{-2}}$:

$$m_2 - m_1 = -\frac{5}{2}\log\left(\frac{I_2}{I_1}\right).$$

(12.9)

Note that if we put in $I_2 = 100 I_1$, then Equation (12.9) gives $m_2 = m_1 - 5$. Thus, if I_2 is 100 times brighter than I_1, then their corresponding magnitudes are such that m_2 is 5 magnitudes *less* than m_1.

Notice that Equation (12.9) only defines magnitudes in terms of their *differences*. We must also agree upon some standard of what represents, for example, $m = 0$. Astronomers define the zero point for the magnitude scale in a complex way that relates directly to the actual process of measuring the brightnesses of stars. Key to this definition are the use of *standard stars*— carefully chosen stars that are easy to observe and seem not to vary in brightness, and so can be used as references of comparison for other stars. The bright naked-eye star Vega in the Northern Hemisphere has played an important historical role in this regard, and so its apparent magnitude is—almost by definition—very close to $m = 0.0$.

We can derive another important relation for magnitudes if we include the inverse square law. Imagine measuring the apparent magnitude of a star, and then moving it ten times farther away and measuring it again. The inverse square law states that the intensity would be less by a factor of $10^2 = 100$. From the definition of the magnitude scale, this means the apparent magnitude of the star would *increase* by 5 magnitudes if one were to (magically) increase the star's distance by a factor of 10. Recall also that the *absolute* magnitude, M, of an object is the apparent magnitude it would have if it were exactly 10 pc distant. If we combine these two facts with Equation (12.9), one can show that the following relation between m, M, and d must hold:

$$m - M = 5 \log d - 5, \tag{12.10}$$

where d is the distance to the object, expressed in parsecs. Notice that for $d = 10\,\text{pc}$, Equation (12.10) gives $m - M = 5\log(10) - 5 = 0$; the apparent and absolute magnitudes are, by definition, equal (and so $m - M = 0$) if the distance is 10 pc. The specific quantity $m - M$ is called the *distance modulus* of the object, because of its connection to distance given by Equation (12.10). If it is greater than zero, then the distance to the star is greater than 10 pc; if the distance modulus is less than zero, then the star is closer than 10 pc.

Equation (12.10) means that if we somehow already know any two of the three quantities m, M, and d, then we can simply solve for the third quantity and calculate it directly. An important point, however, is that *we can always directly measure m*. It is the intensity of the object as seen from Earth, and so we simply place a light meter on a telescope and directly measure it.[4] Thus, we can use Equation (12.10) to calculate the distance, d, if the absolute magnitude is already known, or we can do the opposite. If the distance is known by some other method, then we can use (12.10) to calculate M, and thus the object's luminosity.

We can solve Equation (12.10) for the distance, d, as shown in Equation (12.11):

$$d = 10^{0.2(m-M)+1}. \tag{12.11}$$

Thus, assuming the apparent magnitude, m, is known because we directly measured it, we can use Equation (12.10) to calculate the absolute magnitude, M, if we somehow know the distance. Or we can use Equation (12.11) to calculate the distance if we somehow know the object's true brightness—that is, it's absolute magnitude, M.

---

[4]Of course to *precisely* measure the apparent magnitude of an astronomical object is a complex process in practice. But it is straightforward in principle.

Both of these processes are important, and they depend upon each other. It is, of course, circular logic to apply both Equations (12.10) and (12.11) directly to the *same* object. But many astronomical objects such as stars and galaxies are often arranged in *clusters*—groups of objects that are all together in space, and so all located at about the same distance from Earth. And so if we can determine the distance to any one object in the cluster, we get the distances to the other objects for free.

And so, for example, perhaps we see a distant cluster of stars, and within it is a particular type of star that we recognize because of its spectrum of light. And let us imagine that we had previously measured the distance to a nearby example of this exact same type of star by the method of parallax. This means that, for the nearby example with measured distance, we can calculate its absolute magnitude, M, from Equation (12.10); remember that we can always measure m easily. If we now recognize an example of this same type of star in our distant star cluster, we can use its now-known value of M to calculate its distance with Equation (12.11).

But since this star is in a cluster—with all of the stars at the *same* distance—we get the distances to the other stars in the cluster for free. And so now, knowing the distances to those other stars, we can calculate their absolute magnitudes with Equation (12.10), and thus learn the luminosities of other types of stars. This is especially important if some of the stars in the cluster happen to be of types with no examples near enough to measure their distances by the method of parallax.

And if some of these types of stars, of now-known absolute magnitude, are luminous enough to see at vast distances, perhaps we can use them to determine the distance to an even-farther-away cluster of stars.[5] And so we would then be able to determine the absolute magnitudes of all of the types of stars in *that* cluster too. And on and on. This process builds up an *astronomical distance hierarchy*; we use measured distances of closer objects in order to determine the distances of more-distant objects.

12.2.9   SCATTERING OF LIGHT

A light ray is *scattered* if, upon interacting with matter, it splits into many rays in random directions. *Rayleigh scattering* is an important example, and it occurs when light interacts with particles, such as individual atoms or molecules, that are much smaller than the wavelength of the light. Rayleigh scattering of sunlight by air molecules is responsible for the blue of the sky on a clear day. As individual photons encounter individual air molecules (N_2 or O_2), a certain percentage of the light will deflect off in a different direction. But *shorter wavelengths scatter by larger angles*.

Since it is the shorter-wavelength blue light that scatters the most, when one looks up at a cloudless daytime sky one sees blue light. This blue sky light comes from the Sun, ultimately,

---

[5]This assumes that we also have some independent method for *recognizing* the particular type of star—for example from its spectrum or time variations in brightness.

Figure 12.8: Wavelength-dependent scattering in an egg-shaped piece of opalescent glass. The photo on the left was taken with the egg against a black background and illuminated from above. We see the mostly blue light that is scattered back toward us. The photo on the right was taken with the egg illuminated from behind. In this case we see the light that was *not* scattered.

but it is light that would have missed you, had it not been deflected in your direction by Rayleigh scattering.

On the other hand, the setting Sun appears noticeably reddish. In this case you are seeing the light that has *not* been scattered out of your line of sight. Since it is the blue light that mostly scatters off in other directions, you see sunlight that has had blue light subtracted from it. Thus, the Sun appears more reddish.

Wavelength-dependent scattering similar to the Rayleigh effect can also be seen in certain kinds of opalescent glass. Figure 12.8 shows two photographs of an opalescent glass egg. When white light (of a mix of wavelengths) shines upon it from above, one sees only the mostly short-wavelength light scattered backward toward the observer. And so it appears blue. But when one looks *through* it at a source of light from behind, it appears reddish; this is the longer-wavelength light that has *not* been scattered out of your line of sight.

12.2.10 SCATTERING BY INTERSTELLAR DUST

The gas between the stars, that sometimes forms visible nebulae such as the Orion Nebula, also comes with *dust*. The name is misleading because these individual solid particles are far smaller than what we normally think of as dust on Earth. Rather, they are more similar in size to particles of smoke.

These dust grains absorb light, but they also scatter it, and it is a form of wavelength-dependent scattering similar to Rayleigh scattering. Short wavelengths of light are scattered more, while long wavelengths are scattered less. This scattering is most prominent in the visible and ultraviolet wavelengths of light; it is negligibly small for radio wavelengths.

Figure 12.9: Blue light from the Pleiades star cluster is scattered toward us by a thin cloud of intervening dust. (Image made with the Aladin sky atlas, DSS2.)

Figure 12.9 shows the famous Pleiades star cluster. The blue wisps seeming to surround the cluster of stars arise from interstellar dust that lies along the line of sight, in the foreground. We see the mostly short-wavelength blue light that is scattered toward us by the dust. In this particular case, the intervening dust is too thin to obviously redden our view of the hot blue stars behind it.

The interstellar dust in our Milky Way, like the gas, is mostly confined to the plane of the Galaxy's disk. Since we are located in the disk, our visible-wavelength view along the plane of the Milky way is heavily obscured by dust; at visible wavelengths we see only our own neighborhood. To see further we must either use radio telescopes, or look in directions up out of the disk of the galaxy.

The scattering of light by dust has some important consequences for our observations of stars and galaxies.

1. The light of distant objects is dimmed, since some of the light that would have reached us is instead scattered away, out of our line of sight, on its way to us. In addition, some of the light is directly absorbed by dust grains. Both of these effects taken together are called *interstellar extinction*, and it depends on the total amount of dust along our line of sight to the object. It is also strongly dependent on the wavelength of the light. All else being equal, interstellar extinction is greater at shorter wavelengths than longer wavelengths. It

also stands to reason that extinction is likely to be greater for more distant objects, or those that lie along the plane of the Milky Way (where the dust is thickest).

2. The light of distant objects is *reddened* by dust. Since shorter wavelengths are dimmed more than longer wavelengths, the overall color of the object is altered. This is not the same as the redshift caused by the Doppler effect; no individual wavelengths are altered. It is simply that more blue light is scattered away than red light, and so the overall color appears more reddish, similar to the effect shown in Figure 12.8.

It is this second point that saves us. If dust only dimmed stars, without also reddening them, we would have no way to know whether a star appears dim because it is far away or whether it is because we are looking through a lot of dust.

For example, an A-type star has a temperature of about 10,000 K, and is easily identifiable by the very strong hydrogen lines in its spectrum (see Sections 12.2.7 and 13.2.3). At that temperature, the star should emit the peak of its spectrum at ultraviolet wavelengths (see Sections 12.2.4). And so what if we see a star that has the absorption lines of an A-type star, but the redder *color* of a much-cooler star? We know we are looking through a lot of dust, and the dust has reddened the color of the star. By carefully measuring the amount of reddening, we can correct for the extinction by dust, and determine the true distance to the star.

The extinction can be determined as a correction, A, to the distance modulus given by Equation (12.10):[6]

$$m - M = 5 \log d - 5 + A. \tag{12.12}$$

Interstellar extinction makes the distance modulus *larger* than it would otherwise be for a star of that same distance; a star of a particular absolute magnitude (M) *appears* fainter (m is larger) than it should. Thus, the extinction correction, A, must be added to take this effect into account. In practice, many measurements of reddening are needed to accurately determine the extinction correction.

12.2.11 VIEWS AT DIFFERENT WAVELENGTHS

Figure 12.10 shows the same region of sky as imaged by light detectors sensitive to very different wavelengths of electromagnetic waves. The upper-left-hand image shows only light at the short-wavelength blue end of the visible part of the spectrum, while that on the upper-right shows only longer-wavelength red light. And so the top two images were taken with detectors sensitive to opposite ends of the visible spectrum. The bottom two images, however, were taken with detector-filter combinations sensitive to infrared light. The bottom left image was sensitive in the range $2-2.5\,\mu\mathrm{m}$, while that on the right was sensitive at longer wavelenghts still—about $65\,\mu\mathrm{m}$.

For all three images, the telescope was pointed toward the so-called North America Nebula, an emission nebula in a part of the sky that coincides with the dust-laden plane of our Milky

---

[6]In practice, measurements of m, M, and A are specified to be as measured through a particular colored filter.

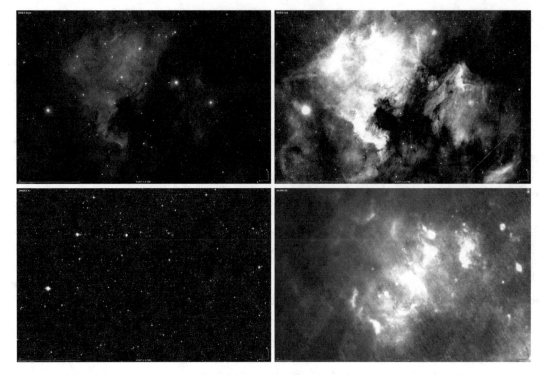

Figure 12.10: Four views of the North America Nebula, at different wavelengths. Clockwise from upper left: blue visible light, red visible light, 65 μm, 2 μm. (Images made with the Aladin sky atlas, DSS2; University of Massachusetts and IPAC/Caltech [Skrutskie et al., 2006]; University of Tokyo, ISAS/JAXA, Tohoku University, University of Tsukuba, RAL and Open University [Doi et al., 2015].)

Way. More stars show in the red-sensitive image, especially in areas that appear relatively dark in the blue-sensitive image. But also, the emission nebula appears much brighter in the red image. The nebula appears brighter because the majority of its light is from the 656.3 nm red emission line of hydrogen gas. And the longer wavelength is less scattered, and so better penetrates the obscuring dust that is the source of the darker patches in the images.

In the 2 μm near-infrared image on the lower left, nearly all traces of the emission nebula are gone, and the longer wavelengths pass easily through the dust, revealing a myriad of stars. At much longer wavelengths of 65 μm shown in the image at lower right, however, the dust itself is seen to glow with thermal radiation, giving an appearance quite unlike the emission-spectrum glowing, ionized *gas* seen in the top two images.

12.3   GRAVITATIONAL WAVES

We have seen that there is a deep connection between the electric and magnetic fields on the one hand, and the phenomenon of electromagnetic waves on the other. So what about the gravitational field? Do *gravitational waves* arise naturally due to the fundamental nature of the gravitational field? Regarding the *classical* gravitational field of Newton, the answer is both "no" and "sort of." The answer is no because, among other reasons, there is no reference to time in Newton's law of gravity; the field is assumed to apply everywhere all at one instant. And so the classical gravitational field has no room for a traveling wave analogous to the far more complex electromagnetic fields.

But the answer is also "sort-of yes" if we think not about the gravitational field itself, but rather objects with mass that are affected by that field. And so there *are* wave-like phenomena that may occur in a large distributions of many masses, all of which are moving according to the mutual gravity they feel for each other. Unlike electromagnetic waves, these are not waves in the field itself; rather they are a wave that passes through material objects, all of which are moving according to Newton's gravitational field. So it is, indirectly, a wave due to gravity. We consider an example in Section 12.4.

But the much more subtle and mathematically complex gravitational field of Einstein's General Relativity is another matter altogether. In Section 11.1.1, I asked a question. What would happen if, for example, I could quickly move the Sun a bit closer to Earth. Would the Earth react to that change instantaneously? Einstein's answer is no; space and time would not alter instantly at distant regions. The action of mass on the geometry of space and time is local too. But altering spacetime near the Sun (by moving it) alters the spacetime next to it. The geometry of space and time here (and now) cannot change without also affecting the points in spacetime that are nearby. This is the very situation that causes a wave.

Thus, if we were to suddenly move the Sun, a disturbance in the geometry of spacetime—a *gravitational wave*—would propagate outward *at the speed of light*. It takes about 8 min for light to travel from the Sun to Earth, and so the motion of Earth would not be affected instantly. Earth would only react when its own local spacetime had changed, and so it would have to wait for the arrival of the gravitational wave, eight minutes later.

Gravitational waves are very difficult to detect in practice, but it is not impossible. To make an easily detectable gravitational wave, an extreme and violent event—involving an enormous quantity of mass in a small volume of space—is required. Examples are supernova explosions or the merger of two black holes. The LIGO experiment accomplished the first successful detection of gravitational waves only as recently as 2016.

Figure 12.11: The grand design spiral galaxy M 51—the Whirlpool Galaxy. (Image made with the Aladin sky atlas. DSS2; Bonnarel et al. [2000], Lasker et al. [1996].)

12.4 SPIRAL DENSITY WAVES

Figure 12.11 shows the famous Whirlpool Galaxy (M 51). It is a good example of a *grand design spiral galaxy*; two almost perfectly symmetrical spiral arms wind around continuously from the outer edges to the central bulge.

The moniker "grand design" calls attention to the fact that something is producing a pattern on a *global* scale in these galaxies. Ordinary wave phenomena are interesting in that patterns on the large, global scale can occur even though all of the action is on the nearby *local* scale. But for mechanical waves such as a sound wave or a wave on a string, these global patterns are a result of *boundary conditions*. Put a sound wave in a long hollow tube, and a global pattern of alternating large and small vibrations occurs; this is the physical basis for a wind instrument such as a flute. Multiple reflections of the wave, back and forth from the ends of the tube, spread the standing wave pattern along the length of the tube.

For a flute, it is unsurprising that there is a large-scale pattern in the wave, because the sound wave is in a tube; the tube provides the global boundary conditions. But what about a galaxy, which has for an enclosure only the infinite vacuum of space? How do the different parts "know" about each other in order to produce the grand-design, global pattern?

Gravity is the key that makes the dynamics of a galaxy fundamentally different from the kind of action that occurs in, say, a wave on a vibrating violin string. For the string, one small piece can directly affect only its immediate neighbors. But each individual star in a galaxy, although it most strongly affects that which is closest to it, puts gravitational forces on distant stars as well. And so gravity allows distant parts of a galaxy to affect each other directly on a global as well as local scale.

This long-distance action of gravity fundamentally changes the nature of wave phenomena, and allows for the possibility of a *spiral density wave*. As orbiting stars and nebulae approach the density wave, they slow down, and so bunch up slightly, making a denser region. This denser region triggers star formation, and that part of the galaxy lights up. It is the most massive and luminous of these newly-formed stars that provides the most light; a single O star can outshine a million Suns. Since these stars don't live long, they are only found in regions of ongoing star formation. And so the slight increase of density in the spiral wave pattern creates the very high-luminosity stars that make the spiral arms visible.

12.5   PROBABILITY WAVES: QUANTUM PHYSICS

The view of light as an electromagnetic wave is wildly successful, able to predict and explain a wide range of phenomena in precise detail. But by the end of the 19th century, it became increasingly clear that some experiments defied explanation with Maxwell's equations alone. The blackbody spectrum of Section 12.2.3 is a good example.

The solid black line in Figure 12.12 shows the spectrum of a 5000 K blackbody as predicted by the electromagnetic theory of the late 19th century, and it is a catastrophic failure of agreement between theory and experiment. At the dawn of the 20th century Max Planck demonstrated that the observed shape of the blackbody spectrum could be explained only if one assumed that energy came in discrete clumps, or *quanta*. This *Planck quantum hypothesis* was the first step in the development of quantum physics, and the idea was extended by Einstein and others to form the modern concept of the *photon*—a particle of light. And so although we observe the spectrum emitted by a blackbody as a wave phenomenon, the fact that light can act as individual particles is essential to its creation.

The particle and wave descriptions of light are connected by Equation (12.13):

$$E = hf = \frac{hc}{\lambda}. \tag{12.13}$$

Here c is the speed of light and E represents the energy of an *individual* photon when the light is acting as a stream of particles. When the same source of light is acting instead like a wave, it has a frequency, f, and a wavelength, λ. Put another way, a source of light of wavelength λ can be seen as made of individual photons, each of which has energy, E, given by Equation (12.13).

This *particle model of light* is necessary in order to provide a physical basis for the nonthermal spectra described in Section 12.2.5. And so physicists of the early 20th century were

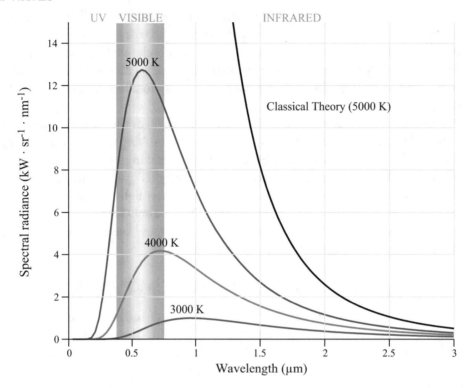

Figure 12.12: The curved marked "classical theory" shows how badly theory fit the data before the development of quantum physics. (Graphic by 4C, CC BY-SA 3.0.)

faced with a dilemma. On the one hand, light clearly does the things that waves do—and so in that sense it *is* a wave. On the other hand, there are many experiments for which a strictly wave explanation fails miserably, but that are explained quite naturally when light is instead assumed to be comprised of discrete photons. And so in these circumstances there is an important sense in which light *is* a stream of particles.

Later evidence showed that this *wave-particle* duality extends to ordinary matter as well as light. We want to think of electrons, for example, as discrete particles of matter. But sometimes their actions can be explained only if they are assumed to act like waves. One of the first successful theoretical explanations for these rather odd observations is the *wave mechanics* of Erwin Schrödinger, published in 1926.

In Schrödinger's theory, even when light (or matter) acts as a stream of particles, there still *is* a wave. This abstract *wave function* is not directly observable. But for any particular well-defined system, it evolves with time in a determined way, governed by the *Schrödinger equation*.

The trick is that although the wave function is not observable, it is related to the *probability*—the likelihood—of a particular observation being made at some place and time. And so we cannot predict, for example, the arrival of an individual photon at a screen—such an event is *random*. But the wave function allows us to calculate the probability of its arrival at that place and time.

These so-called *probability waves* elegantly unify the wave-like and particle-like nature of light and matter, and we take up this subject in more detail in Chapter 14.

12.6 REFERENCES

John Beaver. *The Physics and Art of Photography, Volume 1: Geometry and the Nature of Light.* IOP Publishing, 2018a. DOI: 10.1088/2053-2571/aae1b6 165

John Beaver. *The Physics and Art of Photography, Volume 2: Energy and Color.* IOP Publishing, 2018b. DOI: 10.1088/2053-2571/aae504 174

John Beaver and Charles Conger. Extremely low-cost point-source spectrophotometry (EL-CPSS). *Society for Astronomical Sciences 31st Annual Symposium on Telescope Science*, pp. 113–120, 2012. 171

F. Bonnarel, P. Fernique, O. Bienaymé, D. Egret, F. Genova, M. Louys, F. Ochsenbein, M. Wenger, and J. G. Bartlett. The ALADIN interactive sky atlas. A reference tool for identification of astronomical sources. *Astronomy and Astrophysics Supplement*, 143:33–40, April 2000. DOI: 10.1051/aas:2000331 186

Yasuo Doi, Satoshi Takita, Takafumi Ootsubo, Ko Arimatsu, Masahiro Tanaka, Yoshimi Kitamura, Mitsunobu Kawada, Shuji Matsuura, Takao Nakagawa, Takahiro Morishima, Makoto Hattori, Shinya Komugi, Glenn J. White, Norio Ikeda, Daisuke Kato, Yuji Chinone, Mireya Etxaluze, and Elysandra F. Cypriano. The AKARI far-infrared all-sky survey maps. *Publications of the Astronomical Society of Japan*, 67(3):50, June 2015. DOI: 10.1093/pasj/psv022 184

B. M. Lasker, J. Doggett, B. McLean, C. Sturch, S. Djorgovski, R. R. de Carvalho, and I. N. Reid. The Palomar—ST ScI Digitized Sky Survey (POSS–II): Preliminary Data Availability. In G. H. Jacoby and J. Barnes, Eds., *Astronomical Data Analysis Software and Systems V*, volume 101 of *Astronomical Society of the Pacific Conference Series*, p. 88, 1996. 186

M. F. Skrutskie, R. M. Cutri, R. Stiening, M. D. Weinberg, S. Schneider, J. M. Carpenter, C. Beichman, R. Capps, T. Chester, J. Elias, J. Huchra, J. Liebert, C. Lonsdale, D. G. Monet, S. Price, P. Seitzer, T. Jarrett, J. D. Kirkpatrick, J. E. Gizis, E. Howard, T. Evans, J. Fowler, L. Fullmer, R. Hurt, R. Light, E. L. Kopan, K. A. Marsh, H. L. McCallon, R. Tam, S. Van Dyk, and S. Wheelock. The two micron all sky survey (2MASS). *The Astronomical Journal*, 131:1163–1183, February 2006. DOI: 10.1086/498708 184

CHAPTER 13

Equilibrium

13.1 STATIC EQUILIBRIUM

Equilibrium means that different process are working against each other, but in such a way that some sort of balance is achieved. If it is a *static equilibrium*, then this state of balance remains unchanged over time. If you are sitting in a chair, then you are probably in a state of static equilibrium; the force of gravity downward on you is likely balanced by the force with which the chair pushes upward.

13.1.1 STABILITY AND INSTABILITY

Some forms of static equilibrium are *stable*. This means that if one were to change the situation in such a way that it is no longer in balance, the overall forces are such that they want to bring the system back into equilibrium. An *unstable equilibrium* is the opposite; displace the system slightly from its point of balance, and the tendency is for it to mover *further* away from equilibrium.

There are very familiar examples involving gravity for both stable and unstable equilibrium. Set a round-bottomed bowl on the table and place a marble inside. It is in a stable equilibrium when at the very bottom of the bowl. Nudge the marble in any direction away from the center, and the forces conspire to push it back toward the central equilibrium point.

Now turn the bowl upside down and do the same. There is a point at the very center of the bowl for which all of the forces are balanced on the marble. But clearly, the slightest disturbance from a fly sneezing in the Andromeda galaxy, and the marble rolls faster and faster off the bowl.

Whenever we identify a physical situation that produces static equilibrium, one expects only tiny and very gradual changes to occur. But if the situation instead involves only an unstable equilibrium, or no equilibrium at all, then rapid changes often follow.

13.1.2 HYDROSTATIC EQUILIBRIUM

One of the most important examples of static equilibrium is the balance that takes place at every layer within the Sun and other stars on the main sequence. This balance is called *hydrostatic equilibrium*, and it is an odd sort of name. For the "hydro" in hydrostatic is *water*; and stars are certainly not made of H_2O. But it does make a sort of sense; the same kind of equilibrium is at work both deep in the ocean and deep within the Sun.

Both liquids and gases exert a pressure—a force per unit area—when squeezed. The main difference between the two (so far as hydrostatic equilibrium is concerned) is that gases are easily compressed to smaller volumes whereas liquids are not very compressible. All of the water in the ocean experiences a gravitational force downward. But this downward force of gravity is balanced by the upward force created by the pressure of the water below.

Let us think of a slab of water at some depth in the ocean, with an area, A, in the horizontal direction and a thickness Δy in the vertical direction. It's volume is then simply $V = A\Delta y$, and its mass is its density, ρ, multiplied by its volume. The pressure in the layer of water above our imaginary slab puts a downward force on our slab, while the pressure in the layer below exerts an upward force. But also there is gravity, and it exerts a downward force on the slab of water equal to its weight. *If the slab of water is in hydrostatic equilibrium, then all of these forces must add to zero.*

Given that a force can be seen as a pressure multiplied by the area over which that pressure is exerted, it is easy to show that the following simple relation must hold for *each* layer of water in the ocean:

$$\Delta P = -\rho g \Delta y. \tag{13.1}$$

Here g is the local strength of gravity and ΔP is the *difference* in pressure between the top and bottom of our imaginary slab. The negative sign in Equation (13.1) means that the pressure must *increase* with depth.

Since 13.1 only gives us how the pressure *changes* with depth, to calculate the pressure at a particular location we must add those changes up layer by layer. This takes either simple calculus or complex computing, depending on the complexity of the problem. For water in the ocean, it is relatively simple so long as one does not go too deep, because the pressure does not greatly increase the density of the water.

But the same equation also holds for the inside of the Sun. Only it is the pressure of a gas, not a liquid. This means as one goes deeper into the Sun, the greater pressure will squeeze the gas and increase the density—one of the very things needed to calculate the change in pressure in the first place, and this greatly adds to the complexity of the calculations. Furthermore, the relation between pressure and density for a gas depends also on *temperature*, further complicating the problem.

But all of these complexities can be taken into account in a straightforward way, and just the simple assumption of hydrostatic equilibrium allows one to learn much about the interior of the Sun. It also allows us to ask a question—what if we imagine other stars like the Sun, also in hydrostatic equilibrium, *but with different masses?*

The question turns out to be an important one. With only five, physically rather simple assumptions, much can be learned.

1. The star is in hydrostatic equilibrium.

2. The star is a sphere and begins hydrogen fusion with the same composition as the Sun.

3. The relation between temperature, pressure, and density is given by the simple ideal gas law.

4. Energy is transported outward in the star by either convection or radiation (depending on some easily-tested physical criteria).

5. The star is fusing hydrogen into helium in its core.

One can apply these assumptions to make models of specific stars with different *masses*, spanning the range of known masses of stars. The models will then tell you—given the chosen mass—the visible temperature, radius, and luminosity of the star for each mass. We can then compare these results to observations of real stars.

We can, for example, calculate such models for many different masses and plot their temperatures and luminosities on the H-R diagram. When we do so, the model stars trace out the observed main sequence! More specifically, they lie along the ZAMS. And so we now have a *physical* meaning for the main sequence—these are stars that are in hydrostatic equilibrium and are fusing hydrogen into helium in their cores.

Hydrostatic equilibrium in a main sequence star is an example of a *stable* equilibrium. If one were to try to squeeze the star from the outside somehow, it would make the center of the star hotter. This would make the fusion reactions run faster, which would generate more energy, and that would make the star want to expand. And so our (hypothetical) squeezing is balanced by the very physical consequences of our squeezing.

13.2   DYNAMIC EQUILIBRIUM

There are many processes in the universe whereby different opposing tendencies are balanced, and so it is a kind of equilibrium, even though the opposing processes involve rapid changes in detail. For although the microscopic processes do not balance each other in every detail, they may still average out statistically to a kind of macroscopic balance.

This more subtle form of balance is called a *dynamic equilibrium*. Imagine a popular night club with a long line of individuals waiting to be allowed in, ending at a large bouncer standing at the entrance. The bouncer lets people in only as others exit. And so the names of the individual in the club are constantly changing, but the *number of people in the club remains the same*. The reason for the constant number of people inside is because there is a *dynamic equilibrium between the process of people entering and the process of people leaving.*

Another way to look at the same example is this: if one were to keep record of how many people, on average, entered and left *per minute*, one would find the following: *the rate at which people enter the club is equal to the rate at which people leave the club.*

Although the terms "stable" and "unstable" are not usually applied to the case of a dynamic equilibrium, there is a distinction that has similar importance. A given system may have *feedback mechanisms*—physical properties that increase or decrease the rate of something happening, depending upon the state of the system. And these feedback mechanisms can either drive the system away from equilibrium (positive feedback), or they can help to restore equilibrium (negative feedback) when the system is affected by external forces.

Sometimes, a positive feedback drives the system further and further away from equilibrium until the system changes so much that a new negative feedback occurs that brings the system to a *new* equilibrium—one with very different properties. As an illustration, let us again consider our night club example. At the beginning of the evening, when the doors of the empty club first open, the bouncer intentionally drives the system *away from* equilibrium, allowing the rate of people entering to greatly exceed the rate at which people leave. When the club reaches its legal capacity, the bouncer consciously applies negative feedback—adjusting the rate at which people may enter *in response to the rate at which people are leaving*—in order to establish equilibrium.

What if, on some summer night, the air conditioning cannot keep up with a full house, and it gets too hot in the club? The bouncer decreases the rate at which people may enter to below that at which people are leaving. The night club is then "out of equilibrium" and the number of people decreases with time. When there are few enough people in the club that the air conditioning can keep up, the bouncer establishes a *new* equilibrium by again adjusting the rate at which people enter to match the rate at which people leave.

13.2.1   THERMODYNAMIC EQUILIBRIUM

A hot gas consists of individual particles—some combination of atoms, ions, molecules, and photons (particles of light)—all bouncing off of each other randomly. Each collision involves a transfer of energy from particle to particle, and the particular circumstances are essentially random. The gas is in *thermodynamic equilibrium* when the average rates of transfer of energy between all of the particles, photons included, are equal.

Thermodynamic equilibrium is most easily established if the gas is contained with an "ideal box" that allows no inflow or outflow of energy (see, for example, Carroll and Ostlie [2017, p. 239]). In this state, the physical properties of the gas can be defined by a single temperature, directly proportional to the average energy of the individual particles in the gas. The photons have a range of energies as well, and this results in a spectrum of different wavelengths or frequencies. The spectrum of light for such a gas in thermodynamic equilibrium is the ideal thermal *blackbody spectrum*, described in Section 12.2.3.

Deep inside a star, the gas is, strictly speaking, not in thermodynamic equilibrium. There is a net flow of energy from the center outward, and so the temperature of the gas also changes from layer to layer in the star. But the individual particles in the gas don't "know" about this; the changes in temperature and energy flow are on a much larger scale than the collisions between individual particles. And so we still have a *local thermodynamic equilibrium*, and this means we can describe the gas at any given point in the star as having a particular, well-defined temperature.

13.2.2   ATOMIC LEVEL POPULATIONS

The absorption or emission of spectral lines by a gas is a good example of a dynamic equilibrium. Because of the strange rules of quantum physics, the electrons in an atom can exist only with a specific, predetermined list of energies. It is as if these energies are a set of steps, and the electrons cannot exist with an energy in between the steps. We call these specific steps *energy levels*. The specific energy levels are a property of the atom as a whole—they do not exist for an electron off by itself, not bound by the electrical attraction it feels for the atomic nucleus.

The particular numerical values of these possible energies are different for every atom. Furthermore, they are different for every *ion* of every atom. The energy levels for hydrogen are completely different for those of helium, for example. But also, the energies levels for neutral helium (called He I) are completely different from those of helium that is missing one electron (called He II).

For a given atom or ion, there is a lowest energy level for the electrons, called the *ground state*. All of the other higher energy levels are called *excited states*. Left to itself, an atom most wants to be in the ground state. If it somehow finds itself in an excited state, it usually takes only a fraction of a second before it finds its way back to the ground state.

But there is also a highest possible energy level, called the *ionization energy*. For if the electron is given too much energy, it will escape from the atom altogether, and a new ion results, with a completely new set of energy levels. The energy levels get closer and closer together as they approach this ionization energy. And so although there is a both a lowest and highest energy, and there are only specific values allowed, it is still true that there are infinitely many levels!

Figure 13.1 shows an *energy level diagram* for neutral hydrogen. Only the first four energy levels are shown, and they are numbered, symbolized with the letter n; $n = 1$ is the ground state, and $n = \infty$ (infinity) is the ionization energy. All of the other levels—$n = 5, 6, 7, 8$, etc.—are in between $n = 4$ and $n = \infty$, getting closer and closer to each other as $n = \infty$ is approached.

The energies are labeled in *electron volts (eV)*, a minuscule unit of energy that is more appropriate for the goings-on within individual atoms. Notice that the energies are negative, with the highest energy level equal to zero electron volts. There is nothing odd about this; energy is physically meaningful only in terms of changes.

The arrows between levels in Figure 13.1 represent *transitions*—changes from one energy level to another. In this example, all of the transitions shown are downward—from higher energy to lower energy. But the opposite happens as well (called upward transitions).

Many of the energy levels have more than one possibility, in this case labeled, s, p, d, and f. We will not consider the reasons for this here, but it implies that there are, for example, four distinct ways that a hydrogen atom can be in the $n = 4$ energy state ($E = -0.85\,\text{eV}$), but only one way for it to be in the ground state ($n = 1$, $E = -13.6\,\text{eV}$). This greatly alters the statistics, when an atom changes from one energy to another, and for atoms more complex than hydrogen these levels may even have different energies.

Level Transitions

So how do electrons make upward or downward transitions? There are several ways, but in every case the following must be true: the electron cannot make an upward transition to a higher

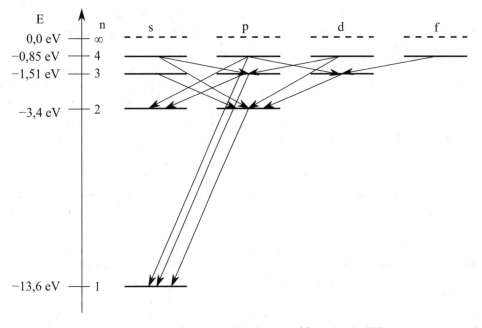

Figure 13.1: An energy level diagram for neutral hydrogen. (Graphic by Wersje rastrowa wykonal uzytkownik polskiego projektu wikipedii: Artura Jana Fijalkowskiego (WarX), Zwektoryzowal: Krzysztof Zajaczkowski, GFDL.)

energy unless some process transfers to it the exact amount of energy gained. Similarly, the electron cannot make a downward transition to a lower energy unless it can transfer the energy lost to some other process. And so what are the processes that can transfer energy to or from an atom, and so allow for upward or downward transitions? The most common of such processes are listed below.

- The atom can collide with free electrons or other atoms in the gas. In the process, collisions can transfer some of the other particle's energy to the atom to provide the energy needed for an upward transition. This is called *collisional excitation*. Similarly, an electron can make a downward transition with the excess energy transferred to the other atom, in a process called *collisional de-excitation*.

- An atom can absorb a particle of light—a photon. The photon disappears, but its energy does not, and it is used to excite the atom to a higher energy level, in a process called *photoexcitation*. This is an all-or-nothing thing; the atom cannot absorb only part of the photon's energy. And so it only happens if the incoming photon has exactly the right amount of energy to bring the electron to one of the possible higher energy levels.

Similarly, the opposite can happen. An electron in an excited state can spontaneously *decrease* its energy, going to a lower energy level, in a process called **spontaneous emission**. In so doing, the excess energy is transformed into a photon of that same energy, emitted by the atom, that travels away at the speed of light. Thus, *an excited atom can emit light*.

An individual photon has energy $E = hf = hc/\lambda$, where h is the tiny Planck constant and f, λ, and c are, respectively, the frequency, wavelength and speed of the light (see Section 12.5). To say that photons of only particular energies can be absorbed or emitted by the atoms is also to say that photons of only particular *wavelengths or frequencies* may be emitted or absorbed.

- An already-ionized atom can re-capture a free electron in a process called **recombination**. The new electron more often than not recombines to an excited state, rather than the ground state.

- An atom can be ionized, with one or more electrons removed entirely, which then become free electrons in the gas. This can happen by the same processes that produced excitation— collisions with free electrons or other atoms and absorption of photons of light. In both cases, the energy transferred must be greater than the energy required to reach the maximum $n = \infty$ level. Ionization by absorbed photons is called **photoionization** and if from a collision it is called **collisional ionization**.

- For all of these processes, transitions need not occur one step at a time. An electron can jump directly from $n = 1$ to $n = 5$, for example.

There are, however so-called *selection rules*, set by the physics of quantum mechanics, that make certain transitions extremely unlikely, or *forbidden*. Quantum physics is required to calculate these *transition probabilities*.

Atomic Level Populations and Collisions

In a given situation, some combination of all of these processes might occur. For example, in a gas atoms are constantly bumping into each other, and at a given temperature, these collisions have an *average* energy. These collisions produce both excitation of atoms and de-excitation of atoms that are already excited. Furthermore, excited atoms undergo spontaneous de-excitation, and emit photons of light.

These events, for a given atom, are random. And so the electron in some particular hydrogen atom, for example, will change rapidly up and down because of collisions and spontaneous emission. Similar things are also happening for all of the other hydrogen atoms in the gas, all of which are experiencing the same overall conditions, but very different individual circumstances at any one moment.

But the *overall* result of all this random jumping up and down the stairs of energy levels is *not* random. For example, because the temperature of the gas dictates the average energy of collisions, transitions to high energy levels are more likely at higher temperatures. And so although each atom has its own experience, there is a collective average. *The atoms in a higher-temperature gas tend to be, on average, at higher levels of excitation than the atoms in a gas at lower temperature.*

Collisional *ionization* is also important. A hydrogen atom, for example, cannot be in the $n = 2$ energy state if it has lost its electron altogether because of a high-energy collision. And recall, for example, that He II (1 electron missing) and He I (neutral) have different energy levels altogether. In general, we find that at higher temperatures, a greater fraction of the atoms have higher levels of ionization as well as excitation.

13.2.3   THE ABSORPTION SPECTRA OF STARS

The gases in the atmosphere of a star are illuminated from a continuous blackbody spectrum coming from deeper layers. Since this is a blackbody spectrum, *all* energies are represented. And so *all* energies of photons pass through the gas. Some of these photons have exactly the correct energy to raise an atom—hydrogen for example—to some other possible energy level. And so these particular photons may be absorbed by the gas, while photons with other energies will not.

The atoms that absorb these photons will then be excited, and they may possibly then spontaneously emit a photon of the exact same energy, and return to the same energy level as before. But this newly-created photon is emitted in a random direction, not the same direction as the photon that excited the atom in the first place. And so there is an overall diminishing of the number of photons that pass through the atmosphere of the star—but only for those photons that have exactly the correct energies to move electrons between energy levels.

Let us consider one particular example, called the *Balmer series of hydrogen*. The Balmer series arises when photons are absorbed that have exactly the right energies to raise hydrogen from the first excited state ($n = 2$) to some higher excited state ($n = 3$ or $4, 5, 6, 7$, etc.). Looking at Figure 13.1, it is clear that much less energy is required for these transitions than if they were starting at the ground state instead; the biggest jump in energy is between $n = 1$ and $n = 2$. But the Balmer series involves transitions in energy that are greater than, for example, any transition from the second excited state ($n = 3$), because higher energy levels are closer to each other.

We can easily calculate the photon energies absorbed for the Balmer series. We simply subtract the $n = 2$ energy from the energies of the different higher levels. And so we conclude that for the Balmer series, photons with these particular energies will be absorbed:

$$-1.51\,\text{eV} - (-3.40\,\text{eV}) = 1.89\,\text{eV}$$
$$-0.85\,\text{eV} - (-3.40\,\text{eV}) = 2.55\,\text{eV}$$
$$-0.54\,\text{eV} - (-3.40\,\text{eV}) = 2.86\,\text{eV}$$
$$-0.38\,\text{eV} - (-3.40\,\text{eV}) = 3.02\,\text{eV}$$
$$-0.28\,\text{eV} - (-3.40\,\text{eV}) = 3.12\,\text{eV}$$
$$-0.21\,\text{eV} - (-3.40\,\text{eV}) = 3.19\,\text{eV}$$
$$-0.17\,\text{eV} - (-3.40\,\text{eV}) = 3.23\,\text{eV}$$
$$-0.14\,\text{eV} - (-3.40\,\text{eV}) = 3.26\,\text{eV}$$
$$-0.11\,\text{eV} - (-3.40\,\text{eV}) = 3.29\,\text{eV}$$
$$-0.09\,\text{eV} - (-3.40\,\text{eV}) = 3.31\,\text{eV}$$
$$\cdots$$
$$0\,\text{eV} - (-3.4\,\text{eV}) = 3.40\,\text{eV}.$$

Here I have added in higher energy levels than those shown in Figure 13.1, up to $n = 14$. I have also put, at the end, the highest possible energy photon that could be absorbed in the Balmer series—3.4 eV—just enough to barely ionize the atom.

In order for a hydrogen atom in the atmosphere of a star to absorb a photon with one of the energies in the Balmer series, *the atom must already be waiting in the $n = 2$ level when the photon arrives*. It is clear, then, that these photons are most likely to be absorbed if a greater fraction of the hydrogen atoms are—at any one moment—excited to $n = 2$. For the conditions within the atmospheres of stars, this happens for hydrogen when the temperature is about 10,000 K.

But what does this have to do with the observed spectra of stars? You may already have guessed that this is the mechanism that produces an *absorption spectrum*, as discussed in Section 12.2.7. These absorbed photon energies relate very directly to the spectrum of the star because *the energy of a photon relates numerically to the wavelength of the light*. The relationship between the two is inverse—so higher photon energies correspond to a *shorter* wavelength of light, according to Equation (12.13). In Table 13.1, I list the calculated wavelengths that are absorbed in the Balmer series of hydrogen. Converting Equation (12.13) to more convenient units, the wavelengths in nanometers can be calculated from the photon energies in electron volts with a simple formula:

$$\lambda = \frac{1240}{E}, \tag{13.2}$$

where E is the photon energy in eV, and λ is the wavelength in nanometers.

The wavelengths absorbed for the Balmer series of hydrogen are in the visible part of the spectrum and the near-ultraviolet, and these are easily observed from ground-based telescopes. Photons absorbed by electrons in the $n = 1$ state of hydrogen, on the other hand, have much higher energy (it is a much bigger jump), and they are far enough into the ultraviolet part of the spectrum that Earth's atmosphere blocks the light. A space telescope is needed to observe

Table 13.1: The Balmer series of hydrogen. When photons are absorbed by an atom, it raises an electron from one energy level (n_1) to another (n_2). The absorbed photon must have exactly the right energy, corresponding to the difference in energy levels (E), in order to do this. Since photon energy is inversely related to the wavelength of light, only certain wavelengths are absorbed. For the Balmer series of hydrogen in particular, it is transitions from the $n = 2$ state to higher energy levels that absorb photons.

n_2	n_1	E(eV)	λ(nm)
3	2	1.89	656
4	2	2.55	486
5	2	2.86	434
6	2	3.02	410
7	2	3.12	397
8	2	3.19	389
9	2	3.23	384
10	2	3.26	380
11	2	3.29	377
12	2	3.31	375
13	2	3.32	374
14	2	3.33	372
∞	2	3.40	365

these spectral lines. Similar concerns apply for photons absorbed by hydrogen atoms in the higher excited states such as $n = 3$—but these wavelengths are only easily observable from space because their wavelengths are too long, in the infrared part of the spectrum.

And so the Balmer lines are important because they are both easily observable and they are from the most abundant element in the universe. Stars near 10,000 K show strong Balmer lines, while stars that are either much hotter or much cooler show weak Balmer lines.

Similar arguments apply to any particular absorption line for any atom one might choose. There is a particular temperature at which it will be the strongest, while it will be weaker for other temperatures. Thus, studying the absorption lines in the spectra of stars is a sensitive test of the temperature of the star's atmosphere, and it leads to the spectral types O, B, A, F, G, K, and M, and the temperature scale associated with them.

Figure 13.2 shows the spectra of stars at different temperatures, covering the range of spectral types. The hot O-type stars are at the top and the cool M-type stars are at the bottom. Notice that each particular absorption line is darkest (relative to the others) at a particular spec-

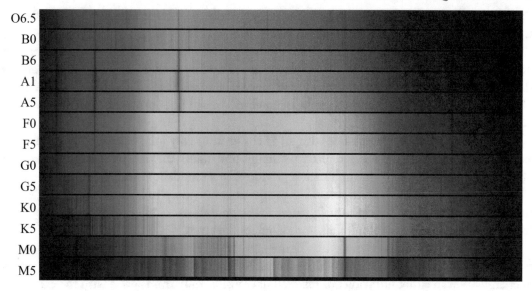

Figure 13.2: Absorption line spectra of different spectral types. The strong dark lines in the A1 star are the hydrogen Balmer lines. (Image credit: NOAO/AURA/NSF, Public Domain.)

tral type. The dark lines prominent for the type A1 star are the Hydrogen Balmer lines. These stars have very nearly the optimal temperature for producing the hydrogen Balmer lines.

One can also see from Figure 13.2 the overall *color* change with temperature. The hottest stars at the top of the figure are much brighter in the blue and violet part of the spectrum than the long-wavelength red end of the spectrum. The opposite is clearly true for the much-cooler M-type stars. The brightest part of the spectrum for a G2 star like the Sun is roughly right in the green middle of the visible spectrum (see Section 12.2.4).

13.2.4   PHOTOIONIZATION-RECOMBINATION EQUILIBRIUM

A gaseous nebula such as the Orion Nebula consists of extremely low-density gas. Left to itself it would be very cold, with nearly all of the atoms in the ground state. And because of the low density, collisions between atoms are much more rare (and much less energetic) than in the atmosphere of a star.

But in the Orion Nebula, the gas is illuminated by the light of the very hot O-type stars that make up the Trapezium. These hot stars emit a considerable amount of ultraviolet light, with photons of energy great enough (E greater than 13.6 eV) to ionize hydrogen atoms, even if they are in the ground state.

And so we have a gas that has been *photoionized*. The electrons can absorb photons with energies somewhat greater than the 13.6 electron volts necessary to free them from the hydrogen nuclei. This extra energy goes into the kinetic energy of motion of the freed electrons. And so the process of photoionization also heats the gas, typically to a temperature of about 10,000 K. Left behind is a hydrogen nucleus without its electron, an H II ion.

While the high-energy photons from the hot Trapezium stars are knocking electrons off neutral hydrogen atoms, previously freed electrons are finding random H II ions and rejoining them to make neutral atoms again; this is the process of *recombination*. For the gas as a whole, photoionization and recombination are in an equilbrium, and both processes occur at the same rate.

When an electron recombines with an H II ion, it usually joins up with the atom in an excited state, rather than joining directly to the ground state. This means that the process of photoionization followed by recombination produces, in a roundabout way, atoms in an excited state. The electrons in these excited states then spontaneously drop to lower energy levels, emitting photons as they go. They eventually arrive at the ground state, unless a high energy photon from the Trapezium comes along first, and reionizes the atom.

And so a photoionized gas emits light—and it emits wavelengths related to all of the possible differences in energy levels of the atom. For hydrogen gas, all of the possible changes in energy result in either ultraviolet or infrared lines visible only from space, with the exception of the Balmer lines.

But in a nebula, these wavelengths are *emitted*, rather than absorbed. And so the gas glows with an emission or bright-line spectrum. Since the brightest of the hydrogen lines is usually red Hα at 656 nm, such an *emission nebula* often appears red in a long-exposure photograph.

13.3   REFERENCES

Bradley W. Carroll and Dale A. Ostlie. *An Introduction to Modern Astrophysics*, 2nd ed., Cambridge University Press, 2017. DOI: 10.1017/9781108380980 194

PART V

Structure

CHAPTER 14

The Structure of Energy and Matter

14.1 THE NATURE OF ENERGY

Energy[1] can take many forms, and it can change from one form to another (or from one form to several others), but it does not disappear, or appear from nothing; the total amount remains constant as stuff happens. This observation is called the *Law of Conservation of Energy*, and it is one of the most important principles in physics.

There is no good one-sentence definition of energy that does not raise more questions than it answers. As with other physical quantities, to be thorough one must use an *operational definition*, describing all of its properties in different circumstances. To do so would require a book of its own; here I give only a brief overview.

There are many types of energy, some of which have familiar names—e.g., electrical energy, solar energy, chemical energy, and nuclear energy. These different types of energy manifest themselves in different ways and under different circumstances (and so they have different names). But when it comes down to it, all types of energy can be described as some variety of four *basic* types, of which I give only very brief descriptions here.

- **Kinetic energy**—This is the energy due to an object's mass and motion, compared to some other frame of reference.

- **Gravitational energy**—This is energy associated with the force of gravity. A glass gets nudged off the table and then falls to the floor and shatters. The energy to make the glass shatter came at the expense of a decrease in gravitational energy.

- **Electromagnetic energy**—The energy associated with electric and magnetic forces is the basis for solar energy and chemical energy, among many other forms. Since light carries electromagnetic energy, it is of particular importance to astronomy.

- **Nuclear energy**—This is the energy associated with the enormous forces at work in the nuclei of atoms.

---

[1]Parts of this chapter appeared, in a somewhat different form, in Beaver [2018b, Chap. 1]

All but the first of these are forms of energy associated with the three fundamental *forces* of nature,[2] and I have listed them in order of the relative strength of these forces. This likely fits with your prior knowledge; compare the power output from an old-fashioned water wheel (gravitational energy of the falling water) to typical sources of electrical power. And then compare these to nuclear energy.

In addition to these explicit forms of energy, there is a direct correspondence between energy and *mass*. Albert Einstein discovered, as part of the theory of Special Relativity, that there is a direct equivalence between mass and energy, embodied in the most famous of all equations:

$$E = mc^2. \tag{14.1}$$

Here c is the speed of light, and Equation (14.1) says that a tiny bit of mass is equivalent to an enormous amount of energy. And so matter can manifest itself as energy and vice versa. This is one of the key discoveries of modern physics, and it is the foundation for many important phenomena, including the fact that the Sun shines.

The joule (J) is the SI unit for energy. As an example of how much energy a joule represents, to lift one gallon of milk from the floor to the kitchen counter top requires an increase of gravitational energy of roughly 40 J. By contrast, the energy content of ordinary matter is nearly unimaginable. From Equation (14.1) a single kilogram of mass is equivalent to about 9×10^{16} J of energy. This is the energy that would be released in the explosion of literally *millions of tons* of chemical explosive. These direct conversions between mass and energy ordinarily occur only in nuclear reactions. For historical images of examples of such enormous releases of energy, see the photography of Michael Light [2013].

Although there are four *basic* forms of energy, there are many other names to describe common cases of energy transfer that are often complex combinations of the four basic types. Below are a few examples of particular importance to astronomy.

1. **Thermal energy:** The individual atoms and molecules in a gas, liquid or solid are constantly in random motions, colliding with each other. And so these molecules—each individually—have kinetic energy. Thermal energy is related to the total kinetic energy of these individual motions. The related concept of *temperature* refers not to this total internal energy, but rather to the average kinetic energy *per particle*.

2. **Solar energy:** Since light is an electromagnetic wave, it carries electromagnetic energy. The intensity of light is related to the rate of energy transfer of electromagnetic energy through a region of space, per square meter.

3. **Chemical energy:** Atoms are comprised of positively charged nuclei and negatively charged electrons, and it is the electrical force between opposite charges that holds an

---

[2]The number of fundamental physical forces in nature depends in part on how one counts. There are, for example, really two fundamental types of nuclear forces, and thus two fundamental types of nuclear energy. I have lumped these together as simply "nuclear energy."

atom together. But when atoms are place near each other, the situation is far more complex, and some of the electrons may feel significant electrical attraction to more than one atomic nucleus at a time, as well as repulsion by the electrons of more than one atom.

The rearrangement of the electrons between neighboring atoms results in a change in electromagnetic energy. Whenever the electrons in atoms rearrange from one configuration to another, the resulting changes in electromagnetic energy are known as *chemical energy*.

14.2   SYMMETRY

If all physical laws can be distilled down to a few simple rules, what are they?[3] When I say "simple," I mean simple when expressed in the appropriate mathematical language—a language that may be as yet unknown, and almost certainly difficult to learn. The search for the answer to this basic question has been one of the primary concerns of theoretical physics over the past 100 years. And key to this search is the concept of *symmetry*.

Symmetry is an idea that is highly intuitive; we seem to know it when we see it. But it has a precise mathematical definition.

Symmetry is an *invariance* (a precisely-defined quantity that remains unchanged) under a *transformation* (a precisely-defined change).

Two simple but important examples are *spherical symmetry* and *axial symmetry*. A sphere is a three-dimensional object; it has length, width, and height. But clearly, there is something about a sphere that makes it only one-dimensional. For it matters not in what direction one travels from the center of a sphere; only distance from the center counts.

We can capture this type of symmetry by defining a *transformation*—a precisely defined set of changes—and what remains *invariant* as we perform those changes. To determine whether a particular object has spherical symmetry, one method (there are others) is as follows:

Rotate the object through *any* arbitrary angle, about *any* axis that passes through its center (this is the transformation). If it remains the same (invariant) after such an arbitrary rotation, then the object has spherical symmetry.

Axial symmetry can be defined in a similar way:

Rotate the object through any angle, about *one particular* axis. If it remains the same (invariant) after these rotations, then the object has axial symmetry about that particular axis.

Axial symmetry is, in a sense, more complex than spherical symmetry, for we must specify the particular axis for the arbitrary rotations. And so axial symmetry, when it occurs, is always in relation to a particular *symmetry axis*. For this reason, when we see an example of axial symmetry

---

[3]Parts of this chapter appeared, in a somewhat different form, in Beaver [2018a, Chap. 10].

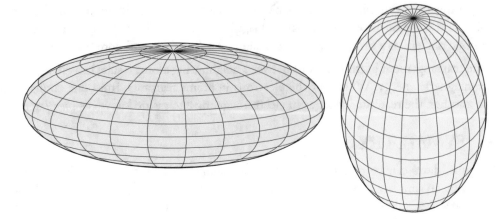

Figure 14.1: Left: oblate spheroid. Right: prolate spheroid. Both types of spheroids have axial symmetry, and both would appear as a simple ellipse when seen at a distance, projected in two dimensions onto the night-time sky. (Graphic by Tomruen, CC BY-SA 4.0.)

in nature an obvious question arises: what is the physical cause of the object's symmetry axis? And why is it *that* axis, and not some other instead?

Elliptical galaxies are a good example of the complications of axial symmetry. They appear roughly as an ellipse in a photograph. An ellipse is a figure that has a high degree of symmetry; it has *mirror symmetry* about a plane passing through either its major or minor axis.

But this view from Earth is only two-dimensional. And so what is the true shape in three dimensions? The simplest three-dimensional shape that appears elliptical is a *spheroid*. One can make a spheroid simply by spinning an ellipse about one of its axes, and so such a figure is also called an *ellipsoid of revolution*; see Figure 14.1. If the symmetry axis lies along the minor axis of the ellipse, it is called an *oblate spheroid*; if instead the symmetry axis is the major axis, it is called a *prolate spheroid*.

Since both oblate and prolate spheroids may appear essentially the same from Earth, then which is it, and why? These are important (and difficult) questions for elliptical galaxies, and we take them up in more detail in Chapter 16. The presence of an axis of symmetry may be a clue to the physical processes at work or it may relate to the body's origin and evolution.

In Section 14.2.1 we look at some other examples of spherical and axial symmetry in astronomy. In Section 14.2.2 we explore the deep connection between symmetry and the fundamental physical laws by which matter and energy interact.

14.2.1   SYMMETRY IN ASTRONOMY

Many astronomical objects exhibit symmetrical forms, and these forms may arise from the nature of the physical processes by which they were made and that govern their evolution. The

Figure 14.2: Spherical symmetry can be seen at vastly different scales of size, when many objects orbit about each other according to the spherically-symmetric gravitational force. From left to right: The Oort cloud of comets surrounding the solar system (about 1 ly across); the globular cluster M 13 (about 170 ly across); the giant elliptical galaxy M 87 (about 1 million light years across). (Image credits—left: ESO/L. Calçada, CC BY-SA 4.0; center: Rawastrodata, CC BY-SA 3.0; right: NASA. STScI, Wikisky, Public Domain.)

gravitational field that arises from a point-like concentration of mass, for example, has *spherical symmetry*. Its direction is always toward the mass, and its strength depends only on the distance. Earth, for example, is not only a point of mass; it is decidedly three dimensional. Nonetheless, gravitation has the overall tendency to pull the matter of Earth into a sphere. This spherical shape of Earth and other self-gravitating astronomical bodies arises from the spherical symmetry of the gravitational force itself.

And it is not just individual self-gravitating bodies that tend toward spherical symmetry. There are many examples of spherical symmetry in *collections* of individual objects. Globular clusters, some elliptical galaxies, the halos of spiral galaxies and the Oort cloud of comets surrounding the solar system are good examples. See Figure 14.2; because of the spherical symmetry of the gravitational force, we often find spherical symmetry at vastly different scales of size.

Even when gravity is the only significant *force* affecting the interaction of objects with each other, it may not be the only *factor*. The self-gravitating body or collection of bodies may, for example, also include an overall *rotation*. Rotation implies an axis, and this may lead to axial symmetry instead of spherical symmetry.

An important ingredient in physical cases of axial symmetry is the *conservation of angular momentum*. The conservation of angular momentum states that there is an overall property of any system that can be calculated, called the total angular momentum. It is related to rotations rather than simple motions, and bodies that are farther from the center (or more massive) count for more than bodies closer to the center of rotation.

This quantity—the total angular momentum of the system—is a quantitative measure of the overall "spinniness" of the system. The conservation of angular momentum states that unless

Figure 14.3: Axial symmetry is seen at vastly different size scales, wherever gravity is the dominant force, but there is also significant rotation present (and thus angular momentum). From left to right: Jupiter is slightly larger in its equatorial direction because of its rotation; artist's conception of the proto-planetary disk that formed the solar system; the Andromeda galaxy. (Image credits—Jupiter: NASA, ESA, Michael Wong (Space Telescope Science Institute, Baltimore, MD), H. B. Hammel (Space Science Institute, Boulder, CO) and the Jupiter Impact Team; Proto-planetary Disk: NASA, Public Domain; M 31: Adam Evans, CC BY 2.0.)

acted upon by an *outside* torque (a kind of twisty force), the total angular momentum of the system remains constant. The conservation of angular momentum is why an ice skater spins faster when they bring their arms inward, and why newly formed neutron stars spin at many hundreds of rotations per second.

Angular momentum has a direction, as well as a magnitude; it points along the axis of rotation. For angular momentum to be conserved, not only its magnitude but also its direction must be preserved. And so the conservation of angular momentum often imposes a symmetry axis on physical processes.

Thus, axial symmetry is commonly seen—at vastly different size scales—in many astronomical bodies or collections of bodies for which an overall rotation is a significant factor. The planet Jupiter, for example, is not really a sphere; it is an oblate spheroid, squished along its rotation axis by its rapid rotation (the same is true, but to a much lesser extent, for Earth). A disk galaxy such as the Milky Way or the Andromeda Galaxy is another obvious example, as is the overall shape of the proto-planetary disk that formed the solar system itself (not including the Oort cloud). See Figure 14.3; as is the case for spherical symmetry, we see these same symmetries at vastly different scales of size.

14.2.2   SYMMETRY AND CONSERVATION LAWS

Symmetry principles are now recognized as some of the most important ways for a physicist to look at the universe. Indeed, physicists often look to symmetry principles in their search for new, as yet undiscovered physical laws, most strikingly in the ongoing search for a *unified field theory* of physics.

It has been known for centuries that there are conserved physical quantities in nature—calculable quantities that stay the same as other measurable quantities change in complex ways. Three important examples are energy, momentum, and angular momentum. The conservation of momentum as a principle was at least partially understood even before Isaac Newton; the conservation of energy was established as a principle by the mid-1800s, and hinted at much earlier.

And so the laws of conservation of energy, momentum, and angular momentum are three examples of eight known conservation laws. Each conservation law specifies a conserved quantity that can be calculated exactly. And even as other measured quantities (velocity, position, acceleration, etc.) change with time in complex ways, the conserved quantity stays the same. These conservation laws are some of the most powerful tools available to a physicist when she applies physical principles to the real world.

If one thinks of a conservation law as a statement that a particular quantity stays the same as everything else changes, it is not too difficult to see that this idea seems related in some way to the concept of symmetry. A symmetry is, after all, an invariance (something that stays the same) under a particular transformation (something that changes).

The two concepts do seem similar, but they are clearly not the same. For one thing, a symmetry involves a *particular* transformation (change), while a conservation law involves a quantity that stays the same even as *all other quantities* change.

But despite this difference, symmetries and conservation laws *are* intimately related, a fact first clearly established by the German mathematician Emmy Noether in 1915 (published in 1918). She proved what is now known as *Noether's First Theorem*, which states that *for every conserved quantity, there is an associated symmetry.*

This means, for example, that if the quantity *energy* is conserved for all physical laws, then there must be some symmetry that all of those physical laws share. Furthermore, she showed how to determine precisely what symmetry is associated with each conservation law. For the example of the conservation of energy, the symmetry is this: the very laws of physics themselves are invariant (remain the same) under the transformation of a translation in time.

Another way to say this is that as time passes, the laws of physics themselves do not change. The inevitable result of this symmetry is that a particular quantity (the total energy of a system) is conserved. The reason then, that all of our physical laws seem to conserve energy, is that they all have this basic symmetry.

Two other important cases are momentum and angular momentum. Both are conserved quantities, and each is associated with a different symmetry. The conservation of momentum arises because—so it appears—the laws of physics do not change from one place to another. Angular momentum, on the other hand, is conserved because the laws of physics do not depend upon which direction one points in space.

When we say that physical laws have symmetries such as invariance under translations in space and time, we are making claims about Nature. But how do we test whether or not these claims are true? Noether's First Theorem shows us the way. The conserved quantities associated with these symmetries are measurable, and so we can measure them in experiments that test whether or not they are conserved. As we build more and more experimental evidence for a particular conservation law, Noether's First Theorem demonstrates that we also provide evidence for that particular symmetry of Nature.

14.3   THE STANDARD MODEL OF PARTICLE PHYSICS

Over the past several decades, a picture has gradually emerged that attempts to bring together all of the known forces and types of matter into one unified scheme. Much of the motivation for the particulars comes from a search for various notions of symmetry acting at the most fundamental level of subatomic particles and the interactions between them.

At this level it is more common to speak of an *interaction* than a force. An interaction can be seen as a particular set of rules by which certain particles affect each other. These interactions are typically studied with large *particle accelerators*, which smack subatomic particles into each other at extremely high energy densities. Two of the largest of these accelerators are at CERN, in Geneva and Fermilab, near Chicago.

But it has been increasingly recognized that these human-made accelerators will never reach energies that are high enough to probe the deepest levels, in order to reach an understanding of all of these interactions. We do, however, live in the aftermath of such a high-powered accelerator experiment—the Big Bang itself. And so increasingly, there is a synergy between particle physics and cosmology; each informs the other.

14.3.1   THE PARTICLES AND THEIR INTERACTIONS

Apart from the familiar protons, neutrons and electrons that make up ordinary matter, and the photons that make up light, there are a myriad of other particles that can be identified—each with its own particular properties—from nuclear reactions and particle accelerator experiments. But it has become increasingly clear that most of these particles are not fundamental themselves; they are combinations of more fundamental particles. This is even true of our familiar friends, the proton and the neutron.

The Standard Model of Elementary Particles makes sense of this seeming chaos of particles, showing how they can all be formed from only three families of matter, along with two groups of interaction particles. The interaction particles produce the forces between particles—and so both matter and the fundamental forces are explained in one framework

A summary of the Standard Model can be seen in Figure 14.4. First, it is separated into *bosons* (red and yellow in the diagram) and *fermions* (purple and green in the figure). Fermions have, at some level, solidity; there is a limit to how tightly packed they can be. Electrons, for example, can only be packed into a density of about $1.4\,M_\odot$ per Earth volume—the maximum

Standard Model of Elementary Particles

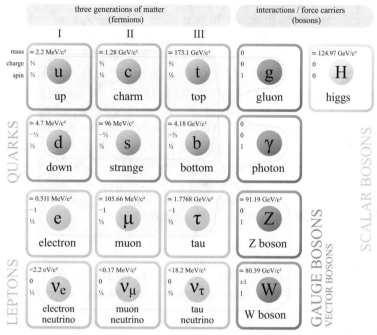

Figure 14.4: A diagram showing the families and types of fundamental particles and their interactions. (Graphic by MissMJ, Public Domain.)

density of a white dwarf. If compressed this tight, they push back, and if forced any tighter, they must change to some other form of particle. This is why a white dwarf can withstand the enormous force of its own gravity without collapsing—it is held up by the enormous pressure of electrons confined to their minimum possible space. A neutron star is similar in that it is also held up by fermions compressed to their smallest possible volume. But for a neutron star it is the much smaller volume of the neutron.

Each of the three families (also called generations) of fermions is broken into two groups: *quarks* and *leptons*. Leptons are fundamental in and of themselves; the most familiar example is the electron. The quarks combine in twos or threes to make other, usually relatively massive particles. The most familiar of these are the proton and neutron, each of which is made of combinations of up and down quarks.

Notice that the most familiar particles—protons, neutrons, and electrons—are all part of the first family. Particles made from combinations of quarks in the second and third families are unstable, and simply do not appear at the low energy densities that occupy our daily lives. But they can be seen in collisions made by particle accelerators. The particles in these three families all have mirror-image particles called *antiparticles*, that share most properties but have (among other things) the opposite electric charge.

Bosons are different from fermions in that there is no restriction to how many of them can occupy a particular volume. The so-called *gauge bosons* transmit forces between the other particles, and control the interactions between them. The photon—a particle of light—is the most familiar example; it is responsible for the electromagnetic force. The *gluon* mediates the forces in the nucleus of an atom, in particular the *strong force* that holds protons and neutrons together to make an atomic nucleus. The *W and Z bosons* were hypothesized in the 1970s to explain the *weak force* that is responsible for certain kinds of radioactive decay. They were finally detected in the 1990s.

Overriding all of these interactions is the *higgs particle*. It was hypothesized by Scottish physicist Peter Higgs in order to help explain why the different particles have the masses that they do. The higgs particle was detected at CERN in 2012; Higgs received the Nobel Prize in Physics the following year, in recognition of that discovery.

14.3.2   WHERE IS GRAVITY?

The Standard Model does an excellent job of unifying, in a relatively simple scheme, much of the complexity of modern physics, regarding the fundamental building blocks of matter and energy and the rules by which they interact with each other. But there is a huge piece missing from it—gravity. Our best theory of gravity is Einstein's GR. But GR is such a different theory in style—it explains gravity in terms of geometry—that it is difficult to know how to fit it in to the Standard Model. There are many ideas and incomplete attempts—probably the most famous is what is known as *string theory*. But the fact remains that we know that we do not know the answer.

Understanding how gravity fits into the Standard Model—or how the Standard Model fits into GR—is necessary if we want to understand the first tiny fraction of a second of the Big Bang. And so pursuit of this ultimate *grand unification* is one of the most active areas of theoretical physics.

14.3.3   WHERE IS DARK MATTER? WHERE IS DARK ENERGY?

The Standard Model contains no clear explanation for dark matter, and has even less to say about dark energy. And so although the Standard Model explains much, there is much that it does not. Why for example does gravity have the particular strength that it does? Why is the universe mostly particles, while antiparticles are rare?

It is possible that these questions are all interconnected, and that they may be unanswerable within the framework of either the Standard Model or GR. Perhaps there is a completely new way of looking at all of this, that naturally brings these ideas together, and thus explains some of what are now mysteries (see, for example, Penrose [2004, Chap. 34]).

14.4   REFERENCES

John Beaver. *The Physics and Art of Photography, Volume 1: Geometry and the Nature of Light*. IOP Publishing, 2018a. DOI: 10.1088/2053-2571/aae1b6 207

John Beaver. *The Physics and Art of Photography, Volume 2: Energy and Color*. IOP Publishing, 2018b. DOI: 10.1088/2053-2571/aae504 205

Michael Light. *100 Suns*. Knopf, New York, 2013. 206

Roger Penrose. *The Road to Reality: A Complete Guide to the Laws of the Universe*. Vintage Books, 2004. 214

CHAPTER 15

The Interior Structure of Stars

15.1 MAIN SEQUENCE STARS

A star is a sphere of gas held together by its own gravity, emitting light because it is very hot. We have already reviewed one of the basic processes at work in a main sequence star: hydrostatic equilibrium. Other processes also apply layer-by-layer within a main sequence star. I summarize them, along with hydrostatic equilibrium.

1. Hydrostatic equilibrium. The pressure within a layer of gas supports the weight of the layers above. This tell us how the pressure, P, varies with distance, r, from the center of the star.

2. The radiation exiting a layer of gas outward is equal to the radiation entering from below, plus any radiation generated within the layer by fusion. This tells us how the luminosity, L, of a layer varies with distance, r, from the center of the star.

3. Heat is transported outward from layer to layer either by conduction (white dwarfs only), radiation (photons of light) or by convection (hot gases rising outward while cooler gases sink inward). This tells us how the temperature, T, varies with distance, r, from the center of the star.

4. The mass of a given layer of gas is equal to the density of the layer times its volume. This tells us how the mass, M, of a layer varies with distance, r, from the center of the star.

5. At any layer within the gas, there is a mathematical relation between temperature, density, and pressure, called the *equation of state*. For many stars, this is simply the *ideal gas law* described in high school chemistry classes.

From each of these principals, one can derive a simple equation that relates how, P, L, M, and T vary within the star, from layer to layer, in terms of r, the distance from the center of the star. More precisely, it tell us the *changes* in these quantities with r. Mathematical analysis of these equations is beyond the scope of this book. But just for fun, I list the four *equations of stellar structure* and the ideal gas law (Equations (15.1)–(15.5)). They correspond to the five descriptions in the list above (I have here listed only one of the three possible equations for energy transport):

$$\frac{dP}{dr} = -\frac{GM\rho}{r^2} \tag{15.1}$$

$$\frac{dL}{dr} = 4\pi r^2 \rho \epsilon \tag{15.2}$$

$$\frac{dT}{dr} = -\frac{3\bar{\kappa}\rho L}{64\pi\sigma r^2 T^3} \tag{15.3}$$

$$\frac{dM}{dr} = 4\pi r^2 \rho \tag{15.4}$$

$$P = \frac{\rho k T}{\mu m_H}. \tag{15.5}$$

We will not solve these equations in this book! Furthermore, there is much other information and data that go with them; there is far more to the innocent-looking little variables ϵ and $\bar{\kappa}$, for example. And notice that the equations are *coupled*—some of the variables appear in more than one equation. These equations must be solved numerically, for each specific example of a star, with a computer. They are solved simultaneously, bit by bit, as one works one's way outward from the center of the star.

A calculated solution to Equations (15.1)–(15.5) makes a particular mathematical model of a star that describes how the pressure, density, temperature, and brightness vary from the center to the star's surface. In the end, the overall mass, luminosity, surface temperature, and radius of the star can be calculated and compared to observations. That the models match up with the observed properties of real stars give us confident in their accuracy. As such, we can then have some faith in what these models say about the not directly observable *inside* of the modeled stars.

One important result of these models is that sometimes heat flows in the star by convection—hot clumps of gas rise outward, carrying their heat with them, while cooler clumps sink inward. This is a very efficient means of transporting heat outward, but there are many instances where it simply doesn't happen. A hot clump is cooled too much if it tries to rise, and so simply sinks back down again.

If convection cannot happen, then the only means to transport heat outward is with radiation—individual photons of light carry their energy as they travel, and so transport energy. But this is a very inefficient process—the photons are mostly smacking into electrons and bouncing around randomly, with the typical distance between collisions only a tiny fraction of a millimeter. And so they only very gradually work their way outward, since they randomly travel inward almost as often as outward.

Conduction, the other possible means of heat transport, is insignificant in ordinary stars; gases simply don't conduct heat very well. But it is the most important mechanism in white dwarfs.

Heat Transfer of Stars

> 1.5 Solar Masses

0.5 − 1.5 Solar Masses

< 0.5 Solar Masses

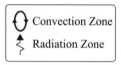
Convection Zone
Radiation Zone

Figure 15.1: Different parts of the interior of a star conduct heat outward by either convection or radiation, depending upon the mass of the star. (Graphic by www.sun.org, CC BY-SA 3.0.)

Figure 15.1 shows the heat transport inside different stars. Some parts transfer heat by convection while other parts by radiation. The exact result depends upon the mass of the star. The middle example includes stars like the Sun. These stars have an inner *radiative zone*, where heat is transported only by radiation, and an outer *convective zone* where heat is transported far more efficiently by convection.

Energy is generated by fusion in only a small *fusion core* in the center. This energy works its way outward very slowly through the radiative zone—taking over a million years. This is striking because it would only take a few seconds if the photons could travel unimpeded by collisions!

Once the heat reaches the convective zone, it is rapidly brought to the visible surface of the star, and this energy then escapes, once again in the form of photons of light—radiation. It is in this form that it makes its way through the transparent vacuum of space to Earth.

For stars greater than about $1.5\,M_\odot$, the situation is reversed; there is a convective zone at the center and a radiative zone on the outside. And the lowest mass stars are convective throughout, with no radiative zone.

Of course from the outside, we don't see these details of heat transfer. But they have significant consequences for the evolution—the life histories—of stars. It is one of the reasons that a low-mass star has a different fate than a high-mass star.

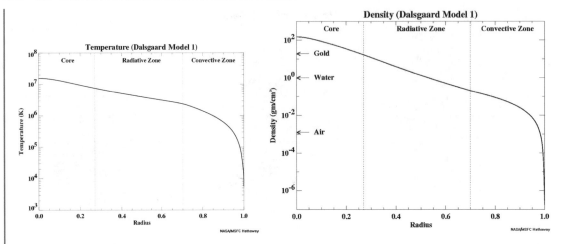

Figure 15.2: Model calculations of the interior temperature (left) and density (right) of the Sun, as a function of distance from the center. (Graphics by Marshall Space Flight Center, NASA/MSFC Hathaway, Public Domain.)

Figure 15.2 shows the result of such a model for the interior of the Sun. The left-hand diagram plots the temperature of the Sun's interior, with the center at the left, and its visible surface at the right. The right-hand graph does the same for the density. Note that the center of the Sun is far more dense than gold or lead—several times denser even than the densest solid on Earth. Yet it is still a gas, compressed by gravity due to the enormous weight of the layers above.

The radiative zone extends to about 70% of the Sun's interior, with the convective zone taking up the outer 30%. The temperature at the center of the Sun is over 10 million kelvin. Near the visible surface of the Sun—its *photosphere*—the temperature and density rapidly decrease.

For all of the interior of the Sun, the temperature is high enough that the gas is nearly completely ionized—all of the electrons are free, rather than bound to nuclei to form atoms. Thus, the interior of the Sun is a gas consisting of atomic nuclei (combinations of protons and neutrons), free electrons, and photons of light.

All of these particles continuously collide with each other and trade energy, and so everything is in a perfect *local thermodynamic equilibrium* that can be described by a single temperature (for a given relatively small location in the Sun's interior). And the light forms a near-perfect blackbody spectrum.

15.1.1   THE FUSION CORE

The region marked "core" in Figure 15.2, extending from the center to the about 0.25 of the distance to the surface, is the only part of the Sun's interior that is both hot and dense enough to extract significant energy from hydrogen fusion.

It is here that the Sun constantly replaces the energy that is lost by the simple fact that it is shining. The Sun's luminosity describes the rate at which energy escapes into space in the form of light. If it had no way to replace this energy at the same rate, it would not be able to keep its interior hot enough—and thus under enough pressure—to balance itself against gravity. And so it would slowly contract, and there would be no hydrostatic equilibrium.

Before the process of nuclear fusion came to be understood in the 20th century, there was no known physical mechanism to explain how the Sun could generate energy at such an enormous rate over billions of years. A basic outline of fusion in the Sun's core is as follows.

- Light atomic nuclei collide and fuse to form a heavier nucleus. In the center of the Sun the overall effect is that four hydrogen nuclei (they are simply single protons) fuse to form a helium nucleus (two protons combined with two neutrons).

- Fusion can generate energy because, in the process, a small percentage of the mass is turned into energy. Einstein's famous equation $E = mc^2$ tells us that even a tiny amount of mass contains an enormous amount of energy.

- Fusion requires extremely high temperatures and densities because collisions between nuclei have to be strong enough to overcome the enormous electrical repulsion between the positively charged protons in the nuclei.

- There are many ways in which lighter elements can fuse to make heavier elements, but hydrogen fusing to helium is the most efficient way to produce energy by fusion. It is the primary source of energy for all main sequence stars, and it occurs in the central core of the star, where the temperature and density is high enough for fusion to occur.

- Higher temperatures are needed to fuse elements heavier than hydrogen. This is simply because there is a greater electrical repulsion between, for example, two helium nuclei (two protons repelled by two protons) than for two hydrogen nuclei (one proton repelled by one proton).

- Elements heavier than iron can be fused, if the temperature is high enough (it happens in a supernova explosion, for example)—but no energy is generated in the process. For fusion of elements heavier than iron, energy is *consumed* by the process, rather than generated. This means that energy turns into mass, rather than the opposite.

15.2   POST-MAIN SEQUENCE STELLAR STRUCTURE

Since main sequence stars fuse hydrogen to helium in their cores, the percentage of hydrogen in the core is constantly decreasing while that of helium is increasing. Eventually, all of the hydrogen in the fusion core will be exhausted, and at that point the star must change.

It is important to note that since fusion occurs only in the core of the star, it doesn't really matter if there is hydrogen only in *other* parts of the star where it is too cool for fusion to occur, if that hydrogen has no way to make its way to the fusion core.

And this is where convection comes in. Recall that in some parts of stars heat is transported by convection while in other parts it is by radiation. Convection mixes the gas in the star—but if convection cannot occur, then the gas cannot mix. The fusion core of the Sun is in its radiative zone, where no mixing can occur. Thus, as the Sun fuses hydrogen into helium in that core, there is no way to bring in fresh un-fused hydrogen from higher regions.

But for the more massive stars, the opposite is true; the inner part of the star is convective, and so as the smaller fusion core burns hydrogen, new hydrogen is constantly brought in by convection. And so these stars have the entire inner convective zone—much bigger than the fusion core itself—at their disposal for fusion.

The least massive stars are convective throughout their entire interior. And so the core will not run out of hydrogen until the entire star runs out.

15.2.1   GIANTS: HYDROGEN SHELL FUSION

When the lowest-mass stars run out of hydrogen to fuse, it means they have converted the entire star from a mix of hydrogen and helium to pure helium. At this point there is no more energy available form hydrogen fusion and gravity wins. The star, no longer in hydrostatic equilibrium, begins to contract. As it contracts, it gets hotter because gravitational contraction itself is a source of energy (gravitational energy is converted to thermal energy).

But such a star never gets hot enough to fuse the helium that makes up its entire composition. It will eventually contract until it reaches the white dwarf stage. At that point gravity is balanced by the extreme pressure created by the strange *electron degenerate gas* that a white dwarf is squeezed into. Such an object need not be hot to exert this pressure, and so it is stable and gradually cools off.

But more massive stars like the Sun have not fused all of their hydrogen—only that within the fusion core (for stars like the Sun) or in the inner convective zone (for more massive stars). This means there is still fresh hydrogen left in the outer portion of the star.

Initially the same thing happens to the Sun as for the lowest mass stars. Once the hydrogen fuel is gone, gravity begins to squeeze the core, and it contracts and heats up. This drives the inner temperature of the star to much higher temperatures than it had when it was on the main sequence. The gravitational contraction makes regions surrounding the original core hot enough that they can now fuse hydrogen. While on the main sequence these layers were too cool for hydrogen fusion, and so they still are mostly hydrogen.

And so a contracting helium core is surrounded by a shell that is fusing hydrogen. For the outer layers of the star, they see below them now *two* sources of energy—the gravitational contraction of the helium core, and the hydrogen fusion taking place in a shell around the shrinking core. This is *more* than enough energy for the outer layers to balance gravity.

And so the visible-from-the-outside layers of the star expand, and cool in the process. Meanwhile, the hidden core contracts and gets hotter. This *red giant* star is thus getting bigger and cooler on the outside, while the inner core is getting smaller and hotter. The combination of hydrogen shell fusion and core contraction generates more total energy than when the star was on the main sequence, and so the luminosity of the star increases greatly.

For stars considerably more massive than the Sun, the inner layers become hot enough to fuse not only hydrogen, but also helium, and even heavier elements still, making especially carbon, neon and oxygen. The hottest regions are always closer to the center, and so these heaviest fusion products are closer to the center. The lighter elements are fused in shells surrounding the heavier ones. For the most massive stars, an onion-skin like structure eventually builds up, as in Figure 15.3.

15.2.2   END STAGES OF FUSION

For the most massive stars, fusion progresses in the core until iron is formed. This happens in stages; when one source of nuclear fuel is exhausted in the core, the core contracts and gravity heats it up until it can then fuse the product of the previous fusion. Meanwhile, similar things are happening, but steps behind, in successive shells surrounding the core.

When the core fusion finally reaches iron as its end product, there is no more energy source available to the core from fusion. Iron fusion does not turn mass into energy, it turns energy into mass. And so the core of such a star can no longer resist the squeezing of gravity, and it collapses in free fall. This rapid gravitational collapse heats the free-falling matter enormously, and fusion occurs at an accelerating rate, quickly creating all of the elements of the periodic table.

The collapse proceeds until the core has reached the density of a neutron star, at which point it is then held up by the pressure of neutrons. Or if too dense, it collapses to a black hole. In either case, a type II supernova explosion results, and the outer layers of the star are driven off at tens of thousands of kilometers per second.

Thus, fusion in massive stars begins with the hydrogen and helium left over from the Big Bang. But it creates from these two primordial elements all of the rest. The stars then helpfully blast these heavy elements into space. A universe that produced only hydrogen and helium in the Big Bang now has the chemical elements of which important things such as harmonicas, kittens, and beavers are made.

Figure 15.3: In the most massive stars, an onion-skin layering of the products of fusion is built up. The heaviest fusion products are in the center because higher temperatures are required to create them from lighter elements. Once the core is made of iron, there is no more energy available from fusion in order to support the weight of the star. It then collapses catastrophically and causes a type-II supernova explosion. In reality, the layers overlap each other. The diagram is only schematic, and not to scale. (Graphic: R. J. Hall, CC BY 2.5.)

CHAPTER 16

The Structure of Galaxies

A galaxy is a collection of its parts—individual stars, star clusters, gas, and dust. Each piece orbits according to the laws of gravity and motion, moving according to the complex gravitational field of all of the other parts combined. And so the motion of any one star affects the motions of all the others. And yet, out of this unimaginable complexity, an ordered structure often arises.

Edwin Hubble was the first to make a practical classification of the variety of galaxy shapes observed in telescopes. This *Hubble Classification Scheme* [Hubble, 1982] has been expanded over the years since Hubble first proposed it, by Alan Sandage, Gerard de Vaucouleurs, and others [Binney and Merrifield, 1998, Section 4.1.1]. The modern system distinguishes between several possibilities. Does the galaxy have a *disk* with a central *bulge*, is it *irregular* in shape, or does it appear smooth and *elliptical*?

1. If it has a disk with a bulge:

    (a) How big is the bulge, relative to the rest of the disk?

    (b) Are there spiral arms, showing evidence of star-forming activity, with a layer of dust in the center of the disk? Or does the galaxy instead look smooth and uniform?

    (c) Are their many indistinct, tightly wound spiral arms, or just two loose spiral arms making something like the letter "S"?

    (d) Do the spiral arms spiral into the bulge, or instead do they begin at the ends of a central bar that passes through the bulge?

    (e) Is there a ring surrounding the bulge? If so, is the ring in close to the nucleus, or is it way out at the edge?

2. If it is elliptical:

    (a) Is it tiny, huge, or more like the size of typical spiral galaxies?

    (b) How elongated is it? Is it essentially spherical? Or is it stretched out like a long oval?

3. If it is irregular:

    (a) Is it smooth and uniform in texture, with no hints of gas or dust?

    (b) Does it instead show complex knots of gas, dust, and star-forming activity?

    (c) Is it tiny compared to other galaxies?

Figure 16.1: From left to right: M 49, NGC 720, and NGC 3115. The first two are elliptical galaxies (types E2 and E6, respectively), while NGC 3115 is really a lenticular, or S0, galaxy—a type intermediate between elliptical and spiral galaxies. (Images made with the Aladin sky atlas. DSS2; Bonnarel et al. [2000], Lasker et al. [1996].)

One concern is the *morphology* of the galaxy—its apparent shape, either as literally imaged in a telescope or as a best estimate of how it would appear if it could be seen in three dimensions from different angles [Binney and Merrifield, 1998, Chap. 4]. As we shall see, there are patterns to the morphology of galaxies; some features tend to go with others. But there are other concerns apart from morphology. Does the galaxy have an active nucleus, and if so what kind? Is a particular galaxy perhaps a special case of a particular kind of interaction with other galaxies, or might it be a brief happenstance interval in the evolution of an individual galaxy?

16.1   ELLIPTICAL AND LENTICULAR GALAXIES

An ellipse is an oval shape with a precise mathematical curve. Despite the name, a typical *elliptical galaxy* is only roughly in the shape of an ellipse [Binney and Tremaine, 1987, p. 21]. For one thing, there is no obvious edge; such a galaxy fades out, gradually becoming fainter as one looks farther from its center. But for any chosen level of brightness, elliptical galaxies appear with shapes between that of a sphere and an elongated oval, diminishing smoothly in brightness from the center, and with no clear sign of gas or dust.

If we divide the shortest dimension, b, of an elliptical galaxy by its longest, a, we can categorize it by its apparent shape. An elliptical galaxy has type En, where n is a number rounded off from:

$$n = 10 \left(1 - \frac{b}{a} \right). \tag{16.1}$$

And so an E0 galaxy appears circular and has $b/a = 1$, while an E7 galaxy has has a long axis that is over three times its short axis ($b/a = 0.3$). See the first two images in Figure 16.1 for examples of E2 and E6 galaxies.

The most elongated elliptical galaxies are E7. There *are* galaxies more elongated than this, and that otherwise look like ellipticals; they seem to be made entirely of stars and vary smoothly in brightness, with no evidence of gas or dust. But their shape deviates markedly from an ellipse, and the light decreases in brightness from the center according to a different mathematical formula than is the case for ellipticals [Binney and Tremaine, 1987, p. 22].

These *lenticular galaxies* instead show evidence of the beginnings of a disk with a large central bulge. See the right-hand example in Figure 16.1 for an example—the spindle-shaped lenticular galaxy NGC 3115. It apparently consists of a flat disk with a large bulge that is almost but not quite as large as the disk itself. The spindle shape arises because the disk is viewed edge-on from Earth. These lenticular galaxies are classified as S0 in the classification schemes of Hubble and de Vaucouleurs described below.

Elliptical galaxies are classified by their apparent two-dimensional outline as seen at a distance from Earth—not what we guess to be their three-dimensional shapes, if we could see them from all angles. But different three-dimensional shapes may appear with the same two-dimensional outline, when viewed from different vantage points. The oblate and prolate spheroids shown in Figure 14.1 are good examples; both shapes appear as an ellipse when seen from a distance. Because of this ambiguity, the true three-dimensional shapes of elliptical galaxies are largely unknown.

16.2 SPIRAL GALAXIES

Unlike elliptical galaxies, lenticular galaxies are classified according to an estimation from their visual appearance that they are something of a lens shape in three dimensions. This shift in approach continues with spiral galaxies; they are classified not according to their literal appearance, but rather by our best estimation of their shapes in three dimensions, if they could be seen from any angle. All five of the galaxies in Figure 16.2, for example, have nearly the same classification (Sb or Sc spirals; see Section 16.3).

Spiral galaxies show prominent lanes of dust and gas in their disks, most apparent for edge-on views such as that of NGC 4565 in Figure 16.2. The gas is transparent, but the dust shows up as a dark line. Spiral galaxies get their name from a spiral pattern in the disk that emanates from either the central bulge or the ends of a *bar* that passes through the bulge. Galaxies with this bar featured are called *barred spirals*.

A minority of spiral galaxies are called *grand design spirals*; they have two symmetric arms that wind several times around the galaxy. *Flocculent galaxies*, on the other hand, show their spiral pattern with many short, individual features that together add up to a spiral pattern, even though each individual "arm" can only be traced a short distance around the bulge. So-called *multi-arm spiral galaxies* occupy the middle ground between grand-design and flocculent galaxies.

A striking feature of spiral galaxies, both barred and unbarred, is that there is a strong correlation between the character of the spiral arms and the size of the galactic bulge:

> Spirals with a large bulge tend to have tightly-wound, often indistinct, spiral arms. Spirals with a small bulge tend to have prominent spiral arms that are loose and open.

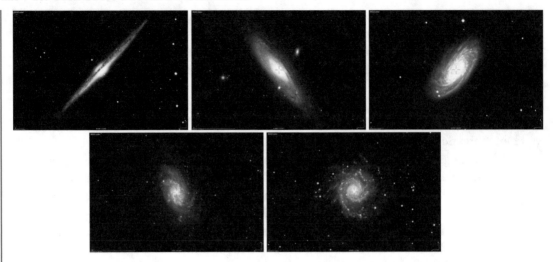

Figure 16.2: A selection of spiral galaxies that appear from Earth at different angles of inclination. Clockwise from upper left: NGC 4565, M 31, M 88, M 74 and M 33. (Images made with the Aladin sky atlas. DSS2; Bonnarel et al. [2000], Lasker et al. [1996].)

This basic fact about spiral galaxies is reflected in the classification schemes described in Sections 16.3 and 16.4.

That stars and star clusters in our own Milky Way can be categorized into separate *populations* (see Chapter 10) applies to other disk galaxies as well. For most spiral galaxies, the gas and dust is strongly concentrated to the plane of the disk, and this is where the bulk of star formation occurs. The halos of disk galaxies, on the other hand, are devoid of the cool, dense clouds of gas and dust needed for star formation.

The visual appearance of a typical spiral galaxy is misleading. Most of the visible light comes from the most luminous stars—those of type O and B. A single O star with 30 times the mass of the Sun can outshine more than 100,000 solar-type stars. These stars are rare, since they don't form as often as lower-mass stars, and they don't last very long. And so we only see them in places where star formation is still occurring or only very recently ceased. HII regions, also associated with ongoing star formation, are highly luminous as well, far outshining their equivalent weight in low-mass stars.

The spiral arms are much brighter than the regions in between them, but there is only a small increase in actual density. But even this increased density triggers star formation, and thus a profound brightening of those slightly denser regions.

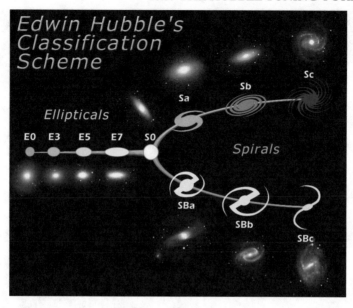

Figure 16.3: The Hubble tuning fork diagram for the classification of galaxy types. (Graphic credit NASA and ESA, CC BY-SA 4.0.)

16.3 THE HUBBLE TUNING-FORK DIAGRAM

The Hubble classification scheme can be arranged into a natural progression of galaxy types, from E0–E7, and then splitting into two branches—one each for unbarred and for barred spirals. Type S0 marks the branching point between ellipticals and spirals in this *tuning fork diagram*; see Figure 16.3.

The spirals seem to make a progression from large bulge with tightly wound spiral arms (Sa) to very small and indistinct bulge with loose, open spiral arms (Sc). Those with a bar feature are labeled, for example, SBb instead of Sb. Taking inspiration from the more-complex de Vaucouleurs classification described in Section 16.4, the Hubble galaxy types are sometimes extended to Sd (or SBd), and irregular galaxies are placed at the ends of the fork.

16.4 THE DE VAUCOULEURS CLASSIFICATION SCHEME

In 1959 Gerard de Vaucouleurs proposed a far more elaborate classification system that accounts for greater variations in the appearance of galaxies. In particular, the de Vaucouleurs classification scheme allows for types intermediate between "normal" and barred spirals. And the system allows for the appearance of ring-like structures in spiral galaxies. Figure 16.4 shows some examples.

Figure 16.4: **Top left:** NGC 3031, type SA(s)ab. **Top center:** NGC 1097, type SB(s)b. **Top right:** NGC 1433, type SB(r)a. **Bottom left:** NGC 1232, type SAB(rs)c. **Bottom center:** NGC 1313, type SB(s)d. **Bottom right:** the same galaxy as bottom center, but imaged in the near infrared instead of blue light. (Images made with the Aladin sky atlas. DSS2; Bonnarel et al. [2000], Lasker et al. [1996].)

The de Vaucouleurs classification uses SA to represent unbarred spirals and SB to denote barred spirals. Intermediate forms—those galaxies with evidence of only a weak bar feature—are labeled SAB. The lowercase letters a, b, and c are used as in the Hubble classification, but a type d is added for galaxies that have essentially no trace of a bulge, but still betray a hint of spiral structure. These d-type spirals are nearly irregular, and so they form a natural morphological bridge between spirals and irregulars. Forms intermediate between, for example, type b and type c, are denoted "bc." Thus, there are really seven possible types: a, ab, b, bc, c, cd, and d.

And finally, the de Vaucouleurs scheme identifies another morphological feature that is absent from the Hubble classification—the presence of a ring-like feature in the disk. As such, the de Vaucouleurs distinguishes between "ring-shaped" and "s-shaped" galaxies, using the letters (r) and (s), in parentheses, to denote the distinction. A galaxy that seems intermediate between the two forms is labeled with (rs).

And so a full de Vaucouleurs classification could be, for example, SA(rs)c—which means an unbarred spiral with a small bulge and loose spiral arms that seems to have a hint of a ring.

The first four galaxies in Figure 16.4 are examples (from left to right and top to bottom), with classifications from de Vaucouleurs [1963].

- **SA(s)ab**: A normal s-shaped spiral (no ring) with a rather large bulge and tightly wound spiral arms, intermediate between Hubble-type Sa and Sb.

- **SB(s)b**: A barred s-shaped spiral with an intermediate-size bulge—equivalent to Hubble type SBb.

- **SB(r)a**: A barred spiral like Hubble type SBa, but with a clear ring connecting the ends of the bar.

- **SAB(rs)c**: Like a Hubble-type Sc galaxy, but with a hint of both a bar and a ring.

The difficulty of deciding between the finely tuned categories in this more-complex scheme is apparent from the lower-left example in Figure 16.4. Is it really midway between a barred and normal spiral? And is it really in between s-shaped and ring-shaped?

The image at bottom center is categorized as a d-type, barred, s-shaped spiral, but it looks like an irregular blob. The image on the bottom right is also of the same galaxy, and the barred spiral shape is more evident. The difference is that the center picture shows the galaxy in blue light, while that on the bottom right shows it in near-infrared light. This highlights another complication of galaxy classification—galaxies often appear quite different depending on the wavelength of the light that is used to image them. Blue light highlights the blue-colored O and B stars that mark regions of star formation; red light de-emphasizes these stars. But red light does emphasize HII regions, which are also associated with star formation. Near-infrared light, on the other hand, tends to suppress both high-luminosity stars and HII regions, better revealing the more-common, lower-mass stars.

16.5   IRREGULAR AND PECULIAR GALAXIES

A galaxy is *irregular* if it cannot be fit neatly in to one of the other morphological categories of elliptical or spiral. But it is not quite so simple as that; the morphology of irregular galaxies is not entirely random. In particular, some irregular galaxies have a hint of spiral structure, especially those with a lot of star formation in regions of gas and dust. The Large Magellanic Cloud, visible from the Southern Hemisphere is a good example (see Figure 2.17). It is often classified as a barred spiral rather than an irregular galaxy. For this reason, irregulars are often categorized as a natural extension of the classification scheme for spiral galaxies. There is no clear evidence, however, that this observed morphology is related to any particular physical or evolutionary cause.

A *dwarf irregular* is a loose aggregation of stars with little central condensation. Some have fewer stars than the largest globular clusters in our own galaxy, although they are much larger in volume. Some dwarf irregulars (the Small Magellanic Cloud is an example) have a lot of gas and dust and show evidence of recent star formation. Others seem to be simply loose aggregations of stars that formed long ago.

A *peculiar* galaxy looks enough like an ordinary galaxy that it can be classified according to the Hubble or de Vaucouleurs scheme. But it is, well, peculiar in some way; it has some odd feature that is atypical of its galaxy type. For many peculiar galaxies, the physical cause of the oddness is apparent; the gravity of a nearby galaxy is distorting its shape.

16.6   THE CAUSES OF GALACTIC STRUCTURE

There is no evidence that the Hubble tuning-fork diagram or the more-complex de Vaucouleurs variant is some kind of evolutionary sequence. Individual galaxies do not gradually change from E0 to E7 and on to Sa through Sd to irregular. So why do galaxies largely fit into such schemes at all? It seems there is no simple answer, and much is unknown. But there are some things that can be said, at least tentatively.

Both theoretical analysis and computer simulations of the dynamics of a disk galaxy—seen simply as a collection of its parts, all moving according to their mutual force of gravity on each other—show that a strong bar structure can arise out of a *dynamical instability*. These *bar instability* modes either appear or they do not. But there are also many dynamical ways to produce weak bars [Binney and Tremaine, 1987]. And so as a first approximation, disk galaxies appear either barred our unbarred (as in the Hubble classification scheme). The more-complex de Vaucouleurs classification scheme, on the other hand, recognizes that intermediate weak bars exist too.

Simulations of spiral density waves have some success in accounting for the structure of grand-design spirals such as M 51 (see Figure 12.11). And the less-organized spiral structure of flocculent galaxies can be explained in a fairly simple way. The shock waves from type II supernovae—associated with massive stars and thus recent star formation—can compress nearby gas so as to trigger more star formation. This self-sustaining star formation can produce bright clumps in the disk of a galaxy. And these brighter star-forming regions can be stretched out into pieces of spiral-like arcs; the inner regions orbit faster than the outer regions.

Many peculiar galaxies clearly owe their oddness to *interactions* with nearby galaxies. The stars in a galaxy orbit due to the gravitational forces arising not only from the combined effect of all the other stars in their own galaxy, but also from the gravitational forces of the stars in the neighboring galaxy. Whenever two large galaxies are very near each other, a host of unusual shapes are bound to result; see Figure 16.5.

Figure 16.5: Two interacting galaxy pairs. Left: NGC 4676. Right: Arp 240. (Images made with the Aladin sky atlas. DSS2; Bonnarel et al. [2000], Lasker et al. [1996].)

16.7   REFERENCES

James Binney and Michael Merrifield. *Galactic Astronomy.* Princeton University Press, 1998. 225, 226

James Binney and Scott Tremaine. *Galactic Dynamics*, vol. 20, Princeton University Press, 1987. DOI: 10.1515/9781400828722 226, 227, 232

F. Bonnarel, P. Fernique, O. Bienaymé, D. Egret, F. Genova, M. Louys, F. Ochsenbein, M. Wenger, and J. G. Bartlett. The ALADIN interactive sky atlas. A reference tool for identification of astronomical sources. *Astronomy and Astrophysics Supplement*, 143:33–40, April 2000. DOI: 10.1051/aas:2000331 226, 228, 230, 233

G. de Vaucouleurs. Revised classification of 1500 bright galaxies. *Astrophysical Journal Supplement*, 8:31, April 1963. DOI: 10.1086/190084 231

Edwin Powell Hubble. *The Realm of the Nebulae.* Yale University Press, 1982. DOI: 10.2307/j.ctt1xp3szq 225

B. M. Lasker, J. Doggett, B. McLean, C. Sturch, S. Djorgovski, R. R. de Carvalho, and I. N. Reid. The Palomar—ST ScI Digitized Sky Survey (POSS–II): Preliminary Data Availability. In G. H. Jacoby and J. Barnes, Eds., *Astronomical Data Analysis Software and Systems V*, volume 101 of *Astronomical Society of the Pacific Conference Series*, p. 88, 1996. 226, 228, 230, 233

CHAPTER 17

Large-Scale Structure of the Universe

17.1 THE Λ-CDM MODEL OF COSMOLOGY

There is a "standard model" of cosmology that has evolved over the past twenty years to something of a tentative consensus. It includes the following parts.

1. A Big Bang, the overall evolution of which is described by Einstein's GR.

2. The standard model of particle physics (see Section 14.3).

3. An inflationary phase of the expansion in the extremely early universe, which guarantees that the universe is exactly flat (Euclidean).

4. A currently accelerating universe, with the acceleration described by a cosmological constant, Λ (dark energy).

5. Dark matter in the form of relatively massive (and thus comparatively low-temperature) weakly-interactive particles. This is called *cold dark matter*, or CDM.

We have already discussed how the first two parts of this basic model explains the existence of the cosmic background radiation and the fact that the universe is made of primarily hydrogen and helium. But the standard model of cosmology can also explain, with some accuracy, the overall large-scale structure of the universe and how it came about.

Figure 17.1 shows two slices through the universe, as measured by the Two-Degree Field Survey of galaxies. Extending to a distance of about 2.5 billion light years, it plots in three dimensions the measured positions of nearly a quarter million galaxies.

The survey clearly shows structures as large as a few hundred million light years across—but nothing bigger than this. At larger distance scales, the universe appears remarkably uniform, in conformance with the cosmological principal. The detail in this diagram shows clusters and superclusters of galaxies, and even larger groupings of superclusters such as the Sloan Great Wall. The galaxy clusters seem to lie along filaments or sheets, with large voids in between.

Figure 17.2 shows a computer simulation of the large scale structure of the universe. It begins with random *quantum fluctuations* in spacetime—a prediction of the quantum physics that underlies the standard model of particle physics. These random fluctuations immediately

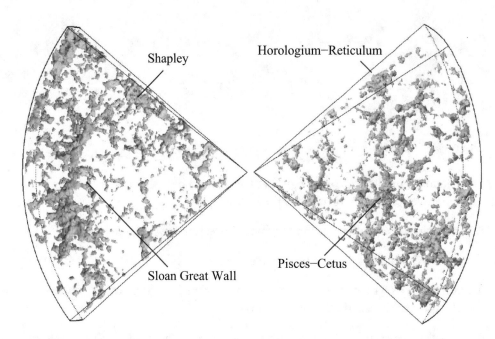

Figure 17.1: The measured positions and distances of nearly a quarter million galaxies from the Two-Degree Redshift Survey are here plotted in three dimensions. Structure at the level of clusters and superclusters is evident as well as even larger structures such as the Sloan Great Wall. The galaxies seem to lie along filaments. (Graphic by Willem Schaap, CC BY-SA 3.0.)

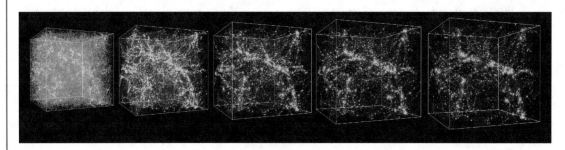

Figure 17.2: A simulation of the development of large-scale structure in the universe using the Λ-CDM standard model of cosmology. (Simulations performed at the National Center for Supercomputer Applications by Andrey Kravtsov (The University of Chicago) and Anatoly Klypin (New Mexico State University). Visualizations by Andrey Kravtsov.)

begin to amplify because of gravity—but only for the CDM. This is crucial because ordinary matter cannot begin to clump significantly until after the time of decoupling. Normal matter is then attracted to the CDM clumps that already exist.

This simulation reproduces the basic features and distance scales of clusters and superclusters of galaxies, such as is shown in Figure 17.1. When we get to finer details—the formation of individual galaxies, for example—the standard model seems to do all right with the rough overall picture. But many of the small details are off.

17.2   ENTROPY AND GRAVITY

A strange feature of our expanding Big Bang universe is that it begins with simplicity and near-perfect uniformity, and becomes *more* complex and *less* uniform over time. This basic and uncontroversial observation about the universe is odd for the following reason: many familiar physical systems tend to do the opposite.

In Section 4.4 we considered *entropy*—the measure of disorder in a physical system. Because there are always a myriad more disordered states than ordered ones, physical systems tend to move, just by the laws of chance, toward states of *more* disorder.

For most physical systems in the laboratory, an increase in entropy comes with an increase in uniformity. Place some gas all in one corner of an evacuated chamber, and it quickly spreads out uniformly throughout. The perfectly uniform spread-out state has *less* order than the original state, with all of the gas concentrated in one place.

But for systems dominated by gravity, an increase in entropy works counter to this expectation. Allow a nearly uniform cloud of gas to pull together under the force of its own self-gravity, and it contracts to a more and more concentrated—less uniform—state. It may seem that this natural process of gravity violates the law of entropy, but it does not. For although the system seen large appears to become more ordered, this is more than made up for at the microscopic level, when one considers the motions of the individual particles of gas. Only gravity does this. The highest entropy state produced by gravity is the maximum concentration of matter—a black hole.

With the natural concentration of gas by gravity comes the inevitable consequence of a rise in temperature, and a greater energy of motion of the individual gas particles. Our Sun formed from the gravitational contraction of a much lower-entropy, uniform cloud of gas at near constant temperature. The action of gravity increased the entropy overall, but with the important consequence of making a hot spot in the universe that emits high-energy photons of light. Although overall the formation of the Sun *increased* the entropy of the universe, the consequence is a source of light that is of very *low* entropy.

This strange relation between gravity and entropy is a good thing! Otherwise, you would not be here to read this book, which wouldn't exist, because I would not be here to have written it. A perfectly uniform universe without structure has no possibility for life, or anything else of interest [Penrose, 2004, Sec. 27.7].

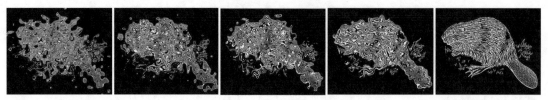

Figure 17.3: When the *flat interaction*, discovered by researchers at Castorium University in Toronto, is used instead of cold dark matter, a very different large-scale structure is predicted for the universe.

17.3   THE FLAT INTERACTION AND LARGE-SCALE STRUCTURE

Very recent, as-yet unpublished research from Castorium University in Toronto, Canada has suggested that there is a hitherto-unknown interaction between quarks and leptons. The Canadian researchers call it the *flat interaction*, and it has an effect something like that of a strong tailwind on a small, single-engine de Havilland. Apparently, it is a very busy force, according to experiments conducted at the Pondwater Accelerator in Bancroft, Ontario. Lead scientist Pat Incisor, announcing the discovery to the media remarked, "Dam! That's for sure a flat interaction!"

The Canadians substituted this interaction for cold dark matter in their simulations, and a very different large-scale structure is predicted for the universe. Unfortunately, these predictions do not compare favorably with observations of clusters and superclusters of galaxies. The simulated structures do, however, match quite well with the morphology of at least one familiar example of Canadian fauna. See Figure 17.3.

17.4   REFERENCES

Roger Penrose. *The Road to Reality: A Complete Guide to the Laws of the Universe*. Vintage Books, 2004. 237

APPENDIX A

Units and Scientific Notation

A.1 UNITS AND DIMENSIONS

When we refer to a physical quantity, it must always have associated with it a set of *dimensions*, and also in many circumstances, a set of *units*.[1] In this context the word "dimension" refers not to spatial dimensions, but rather to the *type* of physical quantity. For example, length is a fundamentally different type of quantity than time. One cannot add a length to a time, nor can one subtract one from the other, because that would equal nonsense. Note that this is not the same thing as apples and oranges. Unlike length and time, one *can* add apples and oranges (it equals fruit salad).

But on the other hand, it is just fine to multiply or divide a length by a time. This produces something with different dimensions, that are a combination of the two. For example, if one divides a length by a time, the result is something that has dimensions of length/time ("length per time"). Often these combined dimensions have special names. This example of length/time has the special name of velocity or speed. And so any time one divides a length by a time, something with dimensions of length/time results.

But what about the actual numbers one plugs into the calculator in a specific case? What if one has a specific length, and a specific time, and wants to calculate a specific speed? Whenever actual numbers are involved, there must also be *units*.

A length of 12.0345 is ambiguous. Is it 12.0345 meters or 12.0345 furlongs? The meter and the furlong are examples of *units*, which are agreed-upon standards for attaching a numerical value to a particular physical quantity. And so the meter is a unit of the dimension of length, and so is a furlong. One can convert between units of the same dimension, by establishing an equivalence between them. And so 1 meter = 3.280 feet = 39.37 inches = 0.00497 furlongs, etc.

In the physical sciences we mostly use a particular international system of units, called *SI*, which stands for "International System" (in French). The SI unit of length is the meter, while the SI unit of time is the second. Every SI unit has an official abbreviation. The abbreviation for the meter is *m*, and for the second it is *s* (it matters that they are lower-case). Table A.1 lists some common SI units, with their dimensions and official abbreviations.

Just as we can derive new dimensions by multiplying or dividing dimensions by each other (length/time, for example), we can do the same for units. And so we can divide meters by seconds to get a new derived unit, which we write m/s (called "meters per second"). What if we want

---

[1] Parts of this chapter appeared, in a somewhat different form, in Beaver [2018].

Table A.1: Common SI units

Dimension	Unit	Abbreviation
Length	Meter	m
Time	Second	s
Mass	Kilogram	kg
Temperature	Kelvin	K
Force	Newton	N
Energy	Joule	J
Power	Watt	W

to divide m/s by seconds? We can do that just fine, and we get m/s/s = m/s^2 (called "meters per second squared"). Many of the units in Table A.1 are actually derived combinations of other units. For example, the newton is actually a combination of kilograms, meters, and seconds:

$$1\,\text{N} = 1\,\text{kg}\,\frac{\text{m}}{\text{s}^2}. \tag{A.1}$$

These base units can be modified by any one of a number of official prefixes, which then multiplies the unit by some power of 10. These prefixes and their abbreviations are listed in Table A.2, although some are more commonly used than others. For example, "milli" means "$\times 1/1000$." And so a millimeter (abbreviated mm) is one thousandth of a meter.

A.2   SCIENTIFIC NOTATION

We have used scientific notation for the values in Table A.2. Physical quantities in nature can vary by many powers of 10. And so for example the light given off by the Sun, it's power, P, is many times greater than the light given off by a 60 W light bulb:

$$P_{\text{Sun}} = 6670000000000000000000000\,P_{\text{lightbulb}}. \tag{A.2}$$

After the 667, there are 24 zeros there. What if I had mistyped (or you miscounted) and you found 23 zeros instead? Well that number would be *ten times* too small. And so clearly, when dealing with numbers like this, we need a better way. And so we use what is called scientific notation. Written this way, the above equation becomes:

$$P_{\text{Sun}} = 6.67 \times 10^{26}\,P_{\text{lightbulb}}. \tag{A.3}$$

The $\times 10^{26}$ part means, $\times 100000000000000000000000000$. But in practical terms this also means, "take the decimal point in 6.67, and move it 26 places to the right, filling in with zeros as needed."

Table A.2: Prefixes for SI units

Prefix	Abbreviation	Meaning
Femto	f	$\times\ 10^{-15}$
Pico	p	$\times\ 10^{-12}$
Nano	n	$\times\ 10^{-9}$
Micro	μ	$\times\ 10^{-6}$
Milli	m	$\times\ 10^{-3}$
Centi	c	$\times\ 10^{-2}$
Deci	d	$\times\ 10^{-1}$
Hecto	h	$\times\ 10^{2}$
Kilo	k	$\times\ 10^{3}$
Mega	M	$\times\ 10^{6}$
Giga	G	$\times\ 10^{9}$
Tera	T	$\times\ 10^{12}$

Raising something to a negative power means the same thing as dividing 1 by that same thing, but raised to the same *positive* power. For example:

$$27^{-3} = \frac{1}{27^3}. \tag{A.4}$$

And so we can also use negative numbers in scientific notation; it means simply *divide* by the power of 10 instead of multiplying by it. And as with positive powers, we can also express this as a decimal equivalent:

$$3.27 \times 10^{-5} = 3.27 \times \frac{1}{10^5} = \frac{3.27}{10^5} = 0.0000327. \tag{A.5}$$

Here we can see that 3.27×10^{-5} means, "take the decimal place in 3.27 and move it 5 places to the *left*, filling in with zeros as needed."

This has a couple of advantages. For one thing, we can see at a glance the most important part numerically: how many powers of ten. Second, when we write it this way, we don't need the zeros for place holders. And so if I put them there, it means I believe that they are significant.

And so, 6.67×10^{26} and 6.670×10^{26} are not really the same number, although they will both appear the same on a calculator. 6.67×10^{26} could possibly be 6.673×10^{26} or even 6.668×10^{26}. If I do not include any more decimal places, then I am making a statement that, based on my uncertainty in the measurement of that quantity, I have no idea what the value of the next decimal place would be. If, on the other hand, I write 6.670×10^{26} then I am saying that I believe (even if with some uncertainty) that it really is 6.670×10^{26} and not, say, 6.673×10^{26}.

Note that one *could* use scientific notation to write the same number in several different ways. You should verify for yourself that the following is true:

$$9.75 \times 10^7 = 975 \times 10^5 = 0.00975 \times 10^{10} = 97500000000 \times 10^{-3}. \qquad \text{(A.6)}$$

Clearly, the last two possibilities look a bit silly, but we try to avoid even the second version. When using scientific notation, it is customary to pick whatever power of 10 is needed in order to have one and only one digit to the left of the decimal place.

A.3 REFERENCES

John Beaver. *The Physics and Art of Photography, Volume 3: Detectors and the Meaning of Digital.* IOP Publishing, 2018. DOI: 10.1088/2053-2571/aaf0ae 239

Author's Biography

JOHN BEAVER

John Beaver is Professor of Physics and Astronomy at the Fox Cities Campus of University of Wisconsin Oshkosh, where he teaches physics, astronomy, photography, and interdisciplinary courses. He earned his B.S. in physics and astronomy in 1985 from Youngstown State University, and his Ph.D. in astronomy in 1992 from Ohio State University. His published work in astronomy is on the topics of spectrophotometry of comets and gaseous nebulae, and multi-color photometry of star clusters.

Beaver is also a fine-art photographer, having exhibited in many juried competitions, and shared and solo exhibitions in Wisconsin, Ohio, New York, Louisiana, Missouri, Oregon, Minnesota, and Colorado. He has long been involved in art-science collaborations (many with artist Judith Baker Waller) in the classroom, at academic conferences, and in art galleries and planetaria. He is the author of *The Physics and Art of Photography*, published in three volumes by Morgan & Claypool Publishers (San Rafael, CA, 2018, 2019). Some of John Beaver's photography can be seen at www.JohnEBphotography.com

Printed in the United States
by Baker & Taylor Publisher Services